Marc Pollefeys   Luc Van Gool
Andrew Zisserman   Andrew Fitzgibbon (Eds.)

3D Structure from Images – SMILE 2000

Second European Workshop on 3D Structure
from Multiple Images of Large-Scale Environments
Dublin, Irleand, July 1-2, 2000
Revised Papers

With 31 Coloured Figures

Springer

Series Editors

Gerhard Goos, Karlsruhe University, Germany
Juris Hartmanis, Cornell University, NY, USA
Jan van Leeuwen, Utrecht University, The Netherlands

Volume Editors

Marc Pollefeys
Luc Van Gool
Katholieke Universiteit Leuven
Center for Processing of Speech and Images
Kasteelpark Arenberg 10, 3001 Leuven-Heverlee, Belgium
E-mail:{Marc.Pollefeys/Luc.VanGool}@esat.kuleuven.ac.be

Andrew Zisserman
Andrew Fitzgibbon
University of Oxford, Department of Engineering Science
19 Parks Road, Oxford OX1 3PJ, UK
E-mail:{az/awf}@robots.ox.ac.uk

Cataloging-in-Publication Data applied for

Die Deutsche Bibliothek - CIP-Einheitsaufnahme

3D structure from images : revised papers / SMILE 2000, Second
European Workshop on 3D Structure from Multiple Images of Large Scale
Environments, Dublin, Ireland, July 1 - 2, 2000. Marc Pollefeys ...
(ed.). - Berlin ; Heidelberg ; New York ; Barcelona ; Hong Kong ;
London ; Milan ; Paris ; Singapore ; Tokyo : Springer, 2001
  (Lecture notes in computer science ; Vol. 2018)
  ISBN 3-540-41845-8

CR Subject Classification (1998): I.4, I.5.4, I.2.10

ISSN 0302-9743
ISBN 3-540-41845-8 Springer-Verlag Berlin Heidelberg New York

This work is subject to copyright. All rights are reserved, whether the whole or part of the material is
concerned, specifically the rights of translation, reprinting, re-use of illustrations, recitation, broadcasting,
reproduction on microfilms or in any other way, and storage in data banks. Duplication of this publication
or parts thereof is permitted only under the provisions of the German Copyright Law of September 9, 1965,
in its current version, and permission for use must always be obtained from Springer-Verlag. Violations are
liable for prosecution under the German Copyright Law.

Springer-Verlag Berlin Heidelberg New York
a member of BertelsmannSpringer Science+Business Media GmbH

http://www.springer.de

© Springer-Verlag Berlin Heidelberg 2001
Printed in Germany

Typesetting: Camera-ready by author, data conversion by PTP Berlin, Stefan Sossna
Printed on acid-free paper   SPIN 10782230   06/3142   5 4 3 2 1 0

Lecture Notes in Computer
Edited by G. Goos, J. Hartmanis

Springer
*Berlin
Heidelberg
New York
Barcelona
Hong Kong
London
Milan
Paris
Singapore
Tokyo*

Preface

This volume contains the final version of the papers originally presented at the second SMILE workshop *3D Structure from Multiple Images of Large-scale Environments*, which was held on 1-2 July 2000 in conjunction with the *Sixth European Conference in Computer Vision* at Trinity College Dublin.

The subject of the workshop was the visual acquisition of models of the 3D world from images and their application to virtual and augmented reality. Over the last few years tremendous progress has been made in this area. On the one hand important new insights have been obtained resulting in more flexibility and new representations. On the other hand a number of techniques have come to maturity, yielding robust algorithms delivering good results on real image data. Moreover supporting technologies – such as digital cameras, computers, disk storage, and visualization devices – have made things possible that were infeasible just a few years ago.

Opening the workshop was Paul Debevec's invited presentation on *image-based modeling, rendering, and lighting*. He presented a number of techniques for using digital images of real scenes to create 3D models, virtual camera moves, and realistic computer animations. The remainder of the workshop was divided into three sessions: *Computation and Algorithms*, *Visual Scene Representations,* and *Extended Environments*. After each session there was a panel discussion that included all speakers. These panel discussions were organized by Bill Triggs, Marc Pollefeys, and Tomas Pajdla respectively, who introduced the topics and moderated the discussion.

A substantial part of these proceedings are the transcripts of the discussions following each paper and the full panel sessions. These discussions were of very high quality and were an integral part of the workshop.

The papers in these proceedings are organized into three parts corresponding to the three workshop sessions. The papers in the first part discuss different aspects of *Computation and Algorithms*. Different problems of modeling from images are addressed – structure and motion recovery, mosaicing, self-calibration, and stereo. Techniques and concepts that are applied in this context are frame decimation, model selection, linear algebra tools, and progressive refinement. Clearly, many of these concepts can be used to solve other problems. This was one of the topics of the discussion that followed the presentation of the papers.

The papers in the second part deal with *Visual Scene Representations*. Papers here deal with concentric mosaics, voxel coloring, texturing, and augmented reality. In the discussion following the presentation of these papers different types of representation were compared. One of the important observations was that the traditional split between image based and geometry based representations is fading away and that a continuum of possible representations exists in between.

The papers in the last part are concerned with the acquisition of *Extended Environments*. These present methods to deal with large numbers of images, the use of special sensors, and sequential map-building. The discussion concentrated on how visual representations of extended environments can be acquired. One of the conclusions was the importance of omnidirectional sensors for this type of application.

Finally we would like to thank the many people who helped to organize the workshop, and without whom it would not have been possible. The scientific helpers are listed on the following page, but thanks must also go to David Vernon, the chairman of ECCV 2000, for his tremendous help in many areas and for organizing a great conference; to the student helpers at Trinity College and in Leuven and to the K.U. Leuven and the ITEA99002 BEYOND project for acting as sponsors of this workshop.

January 2001   Marc Pollefeys, Luc Van Gool
                                          Andrew Zisserman, Andrew Fitzgibbon

Workshop Organizers

Marc Pollefeys — K.U. Leuven
Luc Van Gool — K.U. Leuven and ETH Zürich
Andrew Zisserman — University of Oxford
Andrew Fitzgibbon — University of Oxford

Program Committee

Paul Debevec
Kostas Daniilidis
Chuck Dyer
Andrew Fitzgibbon
Gudrun Klinker

Reinhard Koch
Tomas Pajdla
Shmuel Peleg
Marc Pollefeys
Steve Seitz

Rick Szeliski
Seth Teller
Bill Triggs
Luc Van Gool
Andrew Zisserman

Discussion Chairs

Bill Triggs

Marc Pollefeys

Tomas Pajdla

Review helpers

Kurt Cornelis

Li Zhang

Student Helpers and Session Transcript

Kurt Cornelis
Josephine Sullivan

Maarten Vergauwen
Joss Knight

Frederic Schaffalitzky

Sponsors

K.U. Leuven
ITEA99002 BEYOND project

Table of Contents

Invited Presentation

Pursuing Reality with Image-Based Modeling, Rendering, and Lighting 1
Paul Debevec

Computations and Algorithms

Frame Decimation for Structure and Motion 17
David Nistér

Stabilizing Image Mosaicing by Model Selection 35
Yasushi Kanazawa and Kenichi Kanatani

On Computing Metric Upgrades of Projective Reconstructions under the
Rectangular Pixel Assumption 52
Jean Ponce

A Progressive Scheme for Stereo Matching 68
Zhengyou Zhang and Ying Shan

Panel Session on Computations and Algorithms 86
Bill Triggs, David Nistér, Kenichi Kanatani, Jean Ponce, and Zhengyou Zhang

Visual Scene Representations

Rendering with Non-uniform Approximate Concentric Mosaics 94
Jinxiang Chai, Sing Bing Kang and Heung-Yeung Shum

Volumetric Warping for Voxel Coloring on an Infinite Domain 109
Gregory G. Slabaugh, Thomas Malzbender and W. Bruce Culbertson

A Compact Model for Viewpoint Dependent Texture Synthesis 124
Alexey Zalesny and Luc Van Gool

Augmented Reality Using Uncalibrated Video Sequences 144
Kurt Cornelis, Marc Pollefeys, Maarten Vergauwen and Luc Van Gool

Panel Session on Visual Scene Representation 161
Marc Pollefeys, Sing Bing Kang, Gregory G. Slabaugh, Kurt Cornelis, and Paul Debevec

Extended Environments

Geometry and Texture from Thousands of Images 170
  J.P. Mellor

VideoPlus: A Method for Capturing the Structure and Appearance of Immersive Environments 187
  Camillo J. Taylor

Eyes from Eyes 204
  Patrick Baker, Robert Pless, Cornelia Fermüller and Yiannis Aloimonos

Sequential Localisation and Map-Building in Computer Vision and Robotics 218
  Andrew J. Davison and Nobuyuki Kita

Panel Session on Extended Environments 235
  Tomas Pajdla, J.P. Mellor, Camillo J. Taylor, Tomas Brodsky, and Andrew J. Davison

Author Index 243

Pursuing Reality with Image-Based Modeling, Rendering, and Lighting

Paul Debevec

Institute for Creative Technologies
University of Southern California
13274 Fiji Way 5th Floor
Marina del Rey, CA 90292 USA
paul@debevec.org
http://www.debevec.org/

Abstract. This paper presents techniques and animations developed from 1991 to 2000 that use digital photographs of the real world to create 3D models, virtual camera moves, and realistic computer animations. In these projects, images are used to determine the structure, appearance, and lighting conditions of the scenes. Early work in recovering geometry (and generating novel views) from silhouettes and stereo correspondence are presented, which motivate Façade, an interactive photogrammetric modeling system that uses geometric primitives to model the scene. Subsequent work has been done to recover lighting and reflectance properties of real scenes, to illuminate synthetic objects with light captured from the real world, and to directly capture reflectance fields of real-world objects and people. The projects presented include The Chevette Project (1991), Immersion 94 (1994), Rouen Revisited (1996), The Campanile Movie (1997), Rendering with Natural Light (1998), Fiat Lux (1999), and the Light Stage (2000).

1 Introduction

A prominent goal in computer graphics has been the pursuit of rendered images that appear just as real as photographs. But while graphics techniques have made incredible advances in the last twenty years, it has remained an extreme challenge to create compellingly realistic imagery. For one thing, creating realistic Computer Graphics (CG) models is a time and talent-intensive task. With most software, the artist must laboriously build a detailed geometric model of the scene, and then specify the reflectance characteristics (color, texture, specularity, and so forth) for each surface, and then design and place all of the scene's lighting. Second, generating photorealistic renderings requires advanced techniques such as radiosity and global illumination, which are both computationally intensive and not, as of today, fully general in simulating light transport within a scene. Image-based modeling and rendering (IBMR) can address both of these issues. With IBMR, both the structure and the appearance of the scene is derived from photographs of the real world - which can not only simplify the modeling task, but when employed judiciously can reproduce the realism present in the real-world photographs.

In this article, I present the particular progression of research in this area that I have been involved with. Image-based modeling and rendering is, at its heart, a mixture of image acquisition, image analysis, and image synthesis – or in other words: of photography, computer vision, and computer graphics. I experimented extensively with photography and computer graphics in high school; and my first class in computer vision came in the Fall of 1989 from Professor Ramesh Jain at the Unversity of Michigan. It was in that class, while writing a correlation-based stereo reconstruction algorithm, that it first seemed clear to me that the three disciplines could naturally work together: photography to acquire the images, computer vision to derive 3D structure and appearance from them, and computer graphics to render novel views of the reconstructed scene.

2 The Chevette Project: Modeling from Silhouettes

The first animation I made using image-based techniques came in the summer of 1991 after I decided to create a three-dimensional computer model of my first car, a 1980 Chevette. It was very important to me that the model be truly evocative of the real car, and I realized that building a traditional CG model from grey polygons would not yield the realism I was after. Instead, I devised a method of building the model from photographs. I parked the car next to a tall building and, with help from my friend Ken Brownfield, took telephoto pictures of the car from the front, the top, the sides, and the back. I digitized the photographs then used image editing software to manually locate the silhouette of the car in each image. I then aligned the images with respect to each other on the faces of a virtual box and wrote a program to use the silhouettes to carve out a voxel sculpture of the car (Fig. 1). The surfaces of the exposed voxels were then colored, depending on which way they were facing, by the pixels in the corresponding images. I then created a 64-frame animation of the Chevette flying across the screen. Although (and perhaps because) the final model had flaws resulting from specularities, missing concavities, and imperfect image registration, the realistic texturing and illumination it inherited from the photographs unequivocally evoked an uncanny sense of the actual vehicle. The model also exhibited a primitive form of view-dependent texture mapping, as it would appear to be textured by the front photograph when viewed from the front, and by the top photograph when viewed from the top, etc. As a result, specular effects such as the moving reflection of the environment in the windshield were to some extent replicated, which helped the model seem considerably more life-like than simple texture-mapped geometry. The animation can be seen at the Chevette Project website (see Fig. 1)

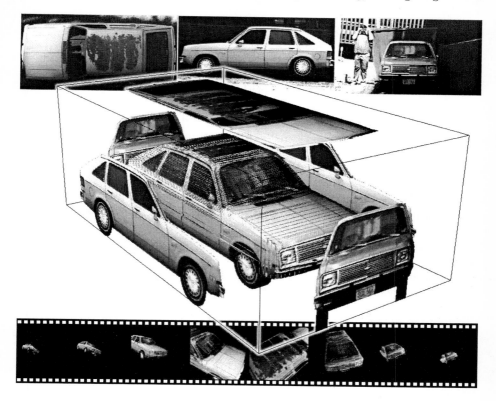

Fig. 1. Images from the 1991 Chevette Modeling project. The top three images show pictures of the 1980 Chevette photographed with a telephoto lens from the top, side, and front. The Chevette was semi-automatically segmented from each image, and these images were then registered with each other approximating the projection as orthographic. The shape of the car is then carved out from the box volume by perpendicularly sweeping each of the three silhouettes like a cookie-cutter through the box volume. The recovered volume (shown inside the box) is then textured-mapped by projecting the original photographs onto it. The bottom of the figure shows a sampling of frames from a synthetic animation of the car flying across the screen, viewable at http://www.debevec.org/Chevette .

3 Immersion '94: Modeling from Stereo

The Chevette project caught the attention of researchers at Interval Research Corporation, where I was hired as a summer intern in the summer of 1994. There I was fortunate to work for Michael Naimark, a media artist who has worked with concepts relating to image-based rendering since the 1970's, and computer vision researcher John Woodfill. Naimark had designed a stereo image capture rig consisting of two Bolex 16mm film cameras fitted with 90-degree-field-of-view lenses eight inches apart atop an aluminum three-wheeled stroller. An encoder attached to one of the wheels caused the cameras to fire synchronously every time the stroller moved a meter forward. In this way, Naimark had filmed several miles of the trails in Banff National Forest.

Our goal for the summer was to turn these sets of stereo image pairs into a photorealistic virtual environment. The technique we used was to determine stereo correspondences, and thus depth, between left-right pairs of images, and then to project the corresponded pixels forward into the 3D world. For this we used a stereo algorithm developed by John Woodill and Ramin Zabih [13]. To create virtual renderings, we projected a supersampled version of the points onto a virtual image plane displaced from the original point of view, using a Z-buffer to resolve the occlusions. Using a single stereo pair, we could realistically re-render the scene from anywhere up to a meter away from the original camera positions, except for artifacts resulting from areas that were unseen in the original images, such as the ground areas behind tree trunks. To fill in the disoccluded areas for novel views, our system would pick the two closest stereo pairs to the desired virtual point of view, and render both to the desired novel point of view. These images were then optically composited so that wherever one lacked information the other would fill it in. In areas where both images had information, the data was linearly blended according to which original view the novel view was closer to - another early form of view-dependent texture mapping. The result was the ability to realistically move through the forest, as long as one kept within about a meter of the original path through the forest. Naimark presented this work at the SIGGRAPH 95 panel "Museums without Walls: New Media for New Museums" [1], and the animations may be seen at the Immersion project website [10].

4 Photogrammetric Modeling with Façade

My thesis work [4] at Berkeley presented a system for modeling and rendering architectural scenes from photographs. Architectural scenes are an interesting case of the general modeling problem since their geometry is typically very structured, and at the same time they are one of the most common types of environment one wishes to to model. The goal of the research was to model architecture in a way that is convenient, requires relatively few photographs, and produces freely navigable and photorealistic results.

The product of this research was Façade [6], an interactive computer program that enables a user to build photorealistic architectural models from a small set of photographs. I began the basic modeling paradigm and user interface at Berkeley in 1993, and later was fortunate to collaborate with Camillo Taylor to adapt his previous work in structure from motion for unorganized line segments [11] to solving for the shape and position of geometric primitives for our project. In Façade, the user builds a 3D model of the scene by specifying a collection of geometric primitives such as boxes, arches, and surfaces of revolution. However, unlike in a traditional modeling program, the user does not need to specify the dimensions or the locations of these pieces. Instead, the user corresponds edges in the model to edges marked in the photographs, and the computer works out

Fig. 2. The Immersion '94 image-based modeling and rendering project. The top images are a stereo pair (reversed for cross-eyed stereo viewing) taken in Banff National Forest. The middle left photo is a stereo disparity map produced by John Woodfill's parallel implementation of the Zabih-Woodfill stereo algorithm [13]. To its right the map has been processed using a left-right consistency check to invalidate regions where running stereo based on the left image and stereo based on the right image did not produce consistent results. Below are two virtual views generated by casting each pixel out into space based on its computed depth estimate, and reprojecting the pixels into novel camera positions. On the left is the result of virtually moving one meter forward, on the right is the result of virtually moving one meter backward. Note the dark disoccluded areas produced by these virtual camera moves; these areas were not seen in the original stereo pair. In the Immersion '94 animations (available at http://www.debevec.org/Immersion , these regions were automatically filled in from neighboring stereo pairs.

the shapes and positions of the primitives that make the model agree with the photographed geometry (Fig. 3).

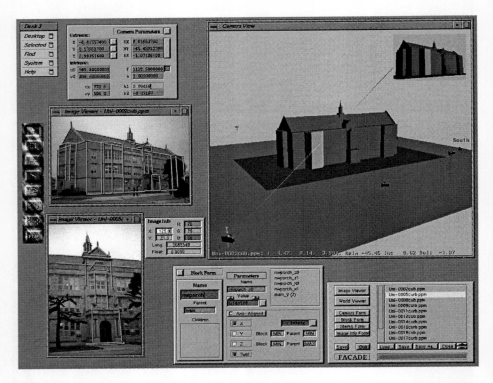

Fig. 3. A screen snapshot from Façade. The windows include the image viewers at the left, where the user marks architectural edge features, and model viewers, where the user instantiates geometric primitives (blocks) and corresponds model edges to image features. Façade's reconstruction feature then determines the camera parameters and position and dimensions of all the blocks that make the model conform to the photographs. The other windows include the toolbar, the camera parameter dialog, the block parameter/constraint dialog, and the main image list window. See also http://www.debevec.org/Thesis/.

Façade simplifies the reconstruction problem by solving directly for the architectural dimensions of the scene: the lengths of walls, the widths of doors, and the heights of roofs, rather than the multitude of vertex coordinates that a standard photogrammetric approach would try to recover. As a result, the reconstruction problem becomes simpler by orders of magnitude, both in computational complexity and, more importantly, in the number of image features that it is necessary for the user to mark. The technique also allows the user to fully exploit architectural symmetries – modeling repeated structures and computing redundant dimensions only once – further simplifying the modeling task.

Like any structure-from-multiple-views algorithm, Façade's reconstruction technique solves for where the original cameras were in the scene. (In addition to the extrinsic position and rotation parameters, Façade is also able to solve for each camera's intrinsic parameters of focal length and center of projection.) With the camera positions known, any one of the photographs can be projected back onto the reconstructed geometry using projective texture mapping. Façade generates photorealistic views of the scene by using all of the available photographs. For each surface point, Façade computes which images it appears in (accounting for visibility), and then blends the pixel values from this set of images to determine the point's appearance in the rendering. This blending can happen in one of several ways. The simple method is to choose entirely the pixel value of the image that viewed the surface point closest to the perpendicular. The more advanced method is to use *view-dependent texture mapping* in which each pixel's contribution to the rendered pixel value is determined as an average weighted by how closely each image's view of the point is aligned with the view of the desired view. As in the Chevette project, blending between the original projected images based on the novel viewpoint helps reproduce some of the effect of specular reflection, but more importantly, it helps simple models appear to have more of the geometric detail present in the real-world scene. With large numbers of original images, the need for accurate geometry decreases, and the VDTM technique behaves as the techniques in the Light Field [9] and Lumigraph [8] image-based rendering work.

Façade was the inspiration for Robert Seidl's photogrammetric modeling product *Canoma*, recently acquired by Adobe Systems from MetaCreations, Inc, and – along with work done at INRIA led by Olivier Faugeras – a source of inspiration for RealViz's ImageModeler software.

Some additional research done in the context of the Façade system enables the computer to automatically refine a basic recovered model to conform to more complicated architectural geometry. The technique, called model-based stereo, displaces the surfaces of the model to make them maximally consistent with their appearance across multiple photographs. Thus, a user can model a bumpy wall as a flat surface, and the computer will compute the relief. This technique was employed in modeling the West façade of the gothic Rouen cathedral for the interactive art installation *Rouen Revisited* shown at the SIGGRAPH 96 art show. Most of the area between the two main towers seen in Fig. 4 was originally modeled as a single polygon. The Rouen project also motivated the addition of new features to Façade to solve for unknown focal lengths and centers of projection in order to make use of historic photographs of the cathedral.

5 The Campanile Movie: Rendering in Real Time

After submitting my thesis at the end of 1996, I continued at Berkeley as a research scientist to create a photorealistic fly-around of the entire Berkeley

Rendering: 1996   Rendering: 1896   Rendering: painting

Fig. 4. Rouen Revisited. Synthetic views of the Rouen cathedral from the *Rouen Revisited* art installation. Left: a synthetic view created from photographs taken in January, 1996. Middle: a synthetic view created from historic postcards showing the cathedral at the time Monet executed his series of paintings (1892-1894). Right: a synthetic view of one of Monet's twenty-eight paintings of the cathedral projected onto its historic geometry, rendering it from a novel viewpoint.

campus. The project took the form of an animated film that would blend live-action video of the campus with computer-rendered aerial imagery, enabling several impossible shifts in perspective. For this project I secured a donation of a graphics computer with hardware texture-mapping from Silicon Graphics, and welcomed graduate students George Borshukov and Yizhou Yu to work on improvements to the rendering and visiblity algorithms in the Façade system.

The main sequence of the film is a swooping fly-around of Berkeley's "Campanile" bell tower, gazing out across the surrounding campus. To create the animation, we built an image-based model of the tower and the surrounding campus – from the foot of the tower out to the horizon – from a set of twenty photographs. I took the photographs from the ground, from the tower, and (thanks to Berkeley professor of architecture Cris Benton) from above the tower using a kite. The final model we built in Façade contained forty of the campus buildings; the buildings further away appeared only as textures projected onto the ground. There were a few thousand polygons in the model, and the sixteen images (Fig. 5) used in rendering the scene fit precisely into the available texture memory of the Silicon Graphics RealityEngine. Using OpenGL and a hardware-accelerated view-dependent texture-mapping technique – selectively blending between the original photographs depending on the user's viewpoint [7] – made it possible to render the scene in real time.

The effect of the animation was one that none of us had seen before – a computer rendering, seemingly indistinguishable from the real scene, able to be

Fig. 5. The Campanile Movie. At top are the original sixteen photographs used for rendering; four additional aerial photographs were used in modeling the campus geometry. In the middle is a rendering of the campus buildings reconstructed from the photographs using Façade; the final model also included photogrammetrically recovered terrain extending out to the horizon. At bottom are two computer renderings of the Berkeley campus model obtained through view-dependent texture mapping from the SIGGRAPH 97 animation. See also http://www.debevec.org/Campanile/.

viewed interactively in any direction and from any position around the tower. The animation, "The Campanile Movie", premiered at the SIGGRAPH 97 Electronic Theater in Los Angeles and would be shown in scores of other venues. Figure 5 shows the model and some renderings from the film. George Borshukov, who worked on the Campanile Movie as a Master's student, went on to join Dan Piponi and Kim Libreri at MANEX Entertainment in applying the Campanile Movie techniques to produce virtual backgrounds for the "bullet-time" shots in the 1999 film *The Matrix* starring Keanu Reeves.

6 Fiat Lux: Adding Objects and Changing Lighting

Façade was used most recently to model and render the interior of St. Peter's Basilica for the animation *Fiat Lux* (Fig. 6), which premiered at the SIGGRAPH 99 Electronic Theater and was featured in the 1999 documentary *The Story of Computer Graphics*. In *Fiat Lux*, our goal was to not only create virtual cinematography of moving through St. Peter's, but to augment the space with animated computer-generated objects in the service of an abstract interpretation of the conflict between Galileo and the church.

The key to making the computer-generated objects appear to be truly present in the scene was to illuminate the CG objects with the actual illumination from the Basilica. To record the illumination we used a high dynamic photography method [5] we had developed in which a series of pictures taken with differing exposures are combined into a *radiance image* – without the technique, cameras do not have nearly the range of brightness values to accurately record the full range of illumination in the real world. We then used an image-based lighting [2] technique to illuminate the CG objects with the images of real light using a global illumination rendering system. In addition, we used an inverse global illumination [12] technique to derive lighting-independent reflectance properties of the floor of St. Peter's, allowing the objects to cast shadows on and appear in reflections in the floor. Having the full range of illumination was additionally useful in producing a variety of realistic effects of cinematography, such as soft focus, glare, vignetting, and lens flare.

7 The Future: Acquiring Reflectance Fields with a Light Stage

In our most recent work we have examined the problem of realistically placing *real* objects into image-based models, taking the photometric interaction of the object with the environment fully into account. To accomplish we have designed a device called a *Light Stage* (Fig. 7) to directly measure how an object transforms incident environmental illumination into reflected radiance, what we refer to as the *reflectance field* of the object. The first version of the light stage consists of a spotlight attached to a two-bar rotation mechanism which can rotate the light in a spherical spiral about the subject in approximately one minute. During

Fig. 6. *Fiat Lux*. The animation *Fiat Lux* shown at the SIGGRAPH 99 Electronic Theater used Façade [6] to model and render the interior of St. Peter's Basilica from single panorama assembled from a set of ten perspective images. Each image was acquired using high dynamic range photography [5], in which each image is taken with a range of different exposure settings and then assembled into a single image that represents the full range of illumination in the scene. This imagery was then used to illuminate the synthetic CG objects which were placed within the scene, giving them the correct shading, shadows, reflections, and highlights. See also http://www.debevec.org/FiatLux/.

this time, one or more digital video cameras record the object's appearance under every form of directional illumination. From this set of data, we can then render the object under any form of complex illumination by computing linear combinations of the color channels of the acquired images as described in [3]. In particular, the illumination can be chosen to be measurements of illumination in the real world [2] or the illumination from a virtual environment, allowing the image of a real person to be photorealistically composited into such a scene with correct illumination. Additional work has been undertaken to render reflectance fields from arbitrary points of view in addition to under arbitrary illumination.

An advantage of this technique for capturing and rendering objects is that the object need not have well-defined surfaces or easy to model reflectance properties. The object can have arbitrary translucency, self-shadowing, interreflection, subsurface scattering, and fine surface detail. This is helpful for modeling and rendering human faces which exhibit all of these properties, as well as for most of the objects that we encounter in our everyday lives.

Fig. 7. Light Stage 1.0. The Light Stage [3] is designed to illuminate an object or a person's face all possible directions in a short period of time. This allows a digital video camera to directly capture the subject's *reflectance field*: how they transform incident illumination into radiant illumination. As a result, we can then synthetically illuminate the subject under any form of complex illumination directly from this captured data. Renderings of synthetically illuminated faces can be found at http://www.debevec.org/Research/LS/ .

8 Conclusion

The advent of image-based techniques have made this an exciting time for research in computer vision and computer graphics, as our ability to model and render aspects of the real world has evolved from approximate models of simple objects to detailed models of complex scenes. Such techniques are already making an impact in the motion picture industry, as image-based modeling, rendering, and lighting has played a role in the most prominent visual effects films of 1999 and 2000. In the next decade we'll be able to capture and display larger data sets, recompute lighting in real time, view scenes as immersive 3D spaces, and populate these recreated spaces with photorealistic digital humans. Some of the most exciting applications of this technology will be for independent filmmakers, as soon it will be possible for a small team of talented people to create a movie with all the visual richness of Star Wars, Titanic, or Lawrence of Arabia, without spending hundreds of millions of dollars – perhaps even opening these techniques for use in education as well as entertainment. What is certain is that image-based techniques will allow us to look forward to a great many new creative visual experiences.

References

1. ADDISON, A. C., MACLEOD, D., MARGOLIS, G., NAIMARK, M., AND SCHWARTZ, H.-P. Museums without walls: New media for new museums. In *Computer Graphics annual Conference Series (SIGGRAPH 95)* (August 1995), R. Cook, Ed., pp. 480–481.
2. DEBEVEC, P. Rendering synthetic objects into real scenes: Bridging traditional and image-based graphics with global illumination and high dynamic range photography. In *SIGGRAPH 98* (July 1998).
3. DEBEVEC, P., HAWKINS, T., TCHOU, C., DUIKER, H.-P., SAROKIN, W., AND SAGAR, M. Acquiring the reflectance field of a human face. *Proceedings of SIGGRAPH 2000* (July 2000), 145–156. ISBN 1-58113-208-5.
4. DEBEVEC, P. E. *Modeling and Rendering Architecture from Photographs*. PhD thesis, University of California at Berkeley, Computer Science Division, Berkeley CA, 1996. http://www.debevec.org/Thesis.
5. DEBEVEC, P. E., AND MALIK, J. Recovering high dynamic range radiance maps from photographs. In *SIGGRAPH 97* (August 1997), pp. 369–378.
6. DEBEVEC, P. E., TAYLOR, C. J., AND MALIK, J. Modeling and rendering architecture from photographs: A hybrid geometry- and image-based approach. In *SIGGRAPH 96* (August 1996), pp. 11–20.
7. DEBEVEC, P. E., YU, Y., AND BORSHUKOV, G. D. Efficient view-dependent image-based rendering with projective texture-mapping. In *9th Eurographics workshop on Rendering* (June 1998), pp. 105–116.
8. GORTLER, S. J., GRZESZCZUK, R., SZELISKI, R., AND COHEN, M. F. The Lumigraph. In *SIGGRAPH 96* (1996), pp. 43–54.
9. LEVOY, M., AND HANRAHAN, P. Light field rendering. In *SIGGRAPH 96* (1996), pp. 31–42.
10. NAIMARK, M., WOODFILL, J., DEBEVEC, P., AND VILLAREAL, L. Immersion '94. http://www.debevec.org/Immersion/, 1994.

11. TAYLOR, C. J., AND KRIEGMAN, D. J. Structure and motion from line segments in multiple images. *IEEE Trans. Pattern Anal. Machine Intell.* (November 1995).
12. YU, Y., DEBEVEC, P., MALIK, J., AND HAWKINS, T. Inverse global illumination: Recovering reflectance models of real scenes from photographs. In *SIGGRAPH 99* (August 1999).
13. ZABIH, R., AND WOODFILL, J. Non-parametric local transforms for computing visual correspondence. In *European Conference on Computer Vision* (May 1994), pp. 151–158.

Discussion

1. **Andrew Fitzgibbon, University of Oxford**: A frivolous question: In "The Matrix", the general appearance is "greeny-grainy", was that colour scheme chosen to simplify the special effects?
    Paul Debevec: That was interesting; they definitely went for a grainy look through the whole film including all of the non-computer graphics non-action shots. That was basically the aesthetic that the Wachowski brothers were going for with their art direction. But as it turns out that's actually a convenient effect for the computer graphics as well, especially since the actors are shot on green screen and often it's difficult to realistically integrate actors shot on green screen into scenes—there is some green spill from the background onto the actors. Of course, there are techniques to get rid of that. I think that the choices made in The Matrix represented a very good marriage of practical limitations and artistic expression. The effects end up looking perhaps a little wrong which totally works in the context of the film's story; the characters are supposed to be in a strange computer-generated world where everything is not quite right.
2. **Hans-Helmut Nagel, Universität Karlsruhe**: Do you have any idea how long it will be before photographic evidence will be banned from court?
    Paul Debevec: Hasn't it been already? I think that any photograph that you bring in is immediately suspect. For example in traffic cases it is quite easy to produce photographic evidence that the stop sign wasn't there. Anyone can perform such fakery with the software that ships with an inexpensive scanner. I would guess photos in criminal cases are scrutinized very heavily, by looking at the original negatives and such. I think video is still used without much question. For example, for the famous video of Rodney King being beaten by the Los Angeles police, nobody questioned whether it was real or not. Today I do not think we could realistically fake such a video even though it was grainy and black-and-white and dark. But I am sure that eventually we will be able to do things like that – probably in five years. It is going to be a matter as much of developing the artistry as of developing the technology. The artists are learning how to make such things happen.
3. **Stefan Heuel, Bonn University**: How long does it take you to acquire 3D models like the campanile or Saint-Peters Basilica?
    Paul Debevec: The first model of the Campanile took me an afternoon to put together. But the version in the film that actually has the arches

and columns was actually built by undergraduates. I had one undergraduate build the lower eighty metres of the tower and the other build the top twenty. It took them about a week, but they were of course learning the system and what the parameters are and things like that. St. Peter's Basilica, I put together in two evenings and the Berkeley campus model was constructed by myself and George Borshukov working about a week each. Much of this time consisted in improving the software which we won't have to do for future projects.

4. **Richard Morris, NASA**: The system you show is very good in terms of computer assisted design. What do you think of more automatic stuff such as techniques from structure from motion?

   Paul Debevec: For the kind of photographic datasets that we have been taking, where the range of different viewpoints is very wide, there is a lot of high-level knowledge that the user gives to the computer that I can't imagine the computer being able to figure out for itself. If you have a view looking down from the tower on the top and a view looking from the side, it is only from our great degree of experience with towers and architectural scenes that we can figure out what corresponds to what. But for systems that use live video as input, there are relatively small motions between the frames and the computer can reasonably figure out what moves to what. We are now seeing some reconstructions from video in, for example, Marc Pollefeys' work [2] that I am very impressed with. It is a little unclear how this could be done if you wanted to perform a reconstruction based on a video of the Berkeley tower. Getting a live video camera up on a kite – that might be difficult. And I think for a pretty wide class of problems (such as building digital sets for movies), it is OK to have it take a while to put the model together. It's usually a very small portion of the total effort on a project; The Campanile Movie involved eight weeks of production of which about a week and a half was putting the model together. So it wasn't on the critical path to getting the film done more quickly, and we had a very fine level of control over the quality of the model, which we needed in making the film look the way we wanted to. So for Hollywood applications there are a lot of things where interactive model-building techniques are going to remain appropriate. But I think there is a whole host of other applications – ones in the film industry as well – that will benefit from the more automatic techniques.

   Andrew Fitzgibbon, University of Oxford: I think one of the interesting messages to the computer vision community—essentially from Paul's Facade work—is to resist the dogma of full automation. There are some cases where manual interaction is useful, and the science remains interesting.

5. **Richard Szeliski, Microsoft**: Now that you are at the new Institute for Creative Technologies, what kind of things do you and the other people in the institute plan to work on?

   Paul Debevec: I'm going to Disneyland! We are basically looking at trying to model very realistic immersive virtual environments. We are going to look into active sensing techniques for that. Basically dealing with large quantities of data. Better quality inverse global illumination for lighting independent models. We are hoping to have some activity in our group that will be looking at the forward rendering problems in global illumination as well. Trying to get those things to be much more efficient – there are several tantalizing systems out there that have solved a part of the global illumination problem nicely such as Hendrik von Jensen's and Eric Veach's system. There is a renderer called "Arnold" that has just been produced by some independents that performs global illumination quite quickly and yields renderings with area light sources which simply no longer look like computer graphics. We want to get some of those things going on. We also want to be able to populate these virtual scenes with people and so some of the work that we did in the waning days of Berkeley was to investigate skin reflectance properties and render virtual faces. We are not animating them yet, but we have some renderings of the faces that seem to have relatively good reflections of the skin, see Debevec *et al* [1]. What we want to do is to get some animated virtual people (hopefully wearing realistic virtual clothing) that can actually go around in these realistic virtual environments ... then we can put something other than big black blocks in them.

References

1. P. Debevec, T. Hawkins, C. Tchou, H.-P. Duiker, W. Sarokin, and M. Sagar. Acquiring the reflectance field of a human face. In *Proceedings, SIGGRAPH*, 2000.
2. M. Pollefeys, R. Koch, M. Vergauwen and L. Van Gool, Metric 3D Surface Reconstruction from Uncalibrated Image Sequences. In *Proc. SMILE Workshop*, R. Koch and L. Van Gool (Eds.), LNCS 1506, pp.138-153, Springer-Verlag, 1998.

Frame Decimation for Structure and Motion

David Nistér

Visual Technology, Ericsson Research
Ericsson Radio Systems
SE-164 80 Stockholm SWEDEN
david.nister@era.ericsson.se

Abstract. A frame decimation scheme is proposed that makes automatic extraction of Structure and Motion (SaM) from handheld sequences more practical. Decimation of the number of frames used for the actual SaM calculations keeps the size of the problem manageable, regardless of the input frame rate. The proposed preprocessor is based upon global motion estimation between frames and a sharpness measure. With these tools, shot boundary detection is first performed followed by the removal of redundant frames. The frame decimation makes it feasible to feed the system with a high frame rate, which in turn avoids loss of connectivity due to matching difficulties. A high input frame rate also enables robust automatic detection of shot boundaries. The development of the preprocessor was prompted by experience with a number of test sequences, acquired directly from a handheld camera. The preprocessor was tested on this material together with a SaM algorithm. The scheme is conceptually simple and still has clear benefits.

1 Introduction

Recently, the Structure and Motion (SaM) branch of computer vision has matured enough to shift some of the interest to building reliable and practical algorithms and systems. The context considered here is the task of recovering camera positions and structure seen in a large number of views of a video sequence. Special interest is devoted to a system that processes video directly from an initially uncalibrated camera, to produce a three-dimensional graphical model completely automatically. Great advances have been made towards this goal and a number of algorithms have been developed [4,8,10,11,15,18,24,26]. However, several additional pieces are necessary for an algorithm to become a full working system and these issues have been relatively neglected in the literature. One such piece, which is proposed here, is a preprocessing mechanism able to produce a sparse but sufficient set of views suitable for SaM. This mechanism has several benefits. The most important benefit is that the relatively expensive SaM processing can be performed on a smaller number of views. Another benefit is that video sequences with different amounts of motion per frame become more isotropic after frame decimation. The SaM system can therefore expect an input motion per frame that is governed by the characteristics of the preprocessor and not by the grabbing frequency or camera movement. Furthermore, problems caused by insufficient motion or bad focus can sometimes be avoided.

The development of the preprocessor was prompted by experience with a number of test sequences, acquired by a non-professional photographer with a handheld camera. It is relatively easy to acquire large amounts of video, which is an important reason for the large interest in structure from motion [1,5,6,7,9,10,12,14,18,20,22,23]. However, obtaining the camera positions in a sequence that covers a lot of ground quickly becomes an awkwardly large problem. Figure 1 has been provided to illustrate the size of the problems that we are interested in. Frame decimation helps reducing the size of these problems.

One way to obtain a smaller set of views is to simply use a lower frame rate than the one produced by the camera. However, this is inadequate for several reasons. First, it can lead to unsharp frames being selected over sharp ones. Second, it typically means that an appropriate frame rate for a particular shot has to be guessed by the user or even worse, predefined by the system. In general, the motion between frames has to be fairly small to allow automatic matching, while significant parallax and large baseline is desirable to get a well-conditioned problem. With high frame rate, an unnecessarily large problem is produced and with low frame rate, the connectivity between frames is jeopardised. In fact, the appropriate frame rate depends on the motion and parallax and can therefore vary over a sequence. Automatic processing can adapt to the motion and avoid any undue assumptions about the input frame rate. Furthermore, unsharp frames caused by bad focus, motion blur etc or series of frames with low interdisparity can be discarded at an early stage. Many algorithms for SaM perform their best on a set of sharp, moderately interspaced still images, rather than on a raw video sequence. A good choice of frames from a video sequence can produce a more appropriate input to these algorithms and thereby improve the final result. In summary, the goal of the preprocessing is to select a minimal subsequence of sharp views from the video sequence, such that correspondence matching still works for all pairs of adjacent frames in the subsequence.

It is possible to identify some desirable properties of a preprocessor. In general, an ideal preprocessor is idempotent. An operator T is called idempotent if $T^2 = T$. In other words, applying the preprocessor twice should yield the same result as applying it once. This is a quality possessed by, for example, ideal histogram equalisation or ideal bandpass filtering. Another desirable property, applicable in this case, is that the algorithm should give similar output at all sufficiently high input frame rates. Furthermore, the algorithm should not significantly affect data that does not need preprocessing.

With large amounts of video, it is rather tedious to start and stop frame grabbing to partition the material into shots. This information should therefore be provided directly from the camera or be derived automatically with image processing. A bonus of being able to handle a high input frame rate is that segmentation of the raw video material into shots can be robustly automated. Automatic detection of shot boundaries can be done rather reliably at high frame rates, while the difference between a discrete swap of camera or view and a large motion diminishes towards lower frame rates. The preprocessing approach is therefore divided into two parts. First, shot boundary detection, which is preferably performed at the output frame rate of the camera. Second, the selection of a subsequence of sharp views representing every shot. The processing is based on a rough global estimation of the rotational camera motion between frames and a sharpness measure. These tools are described in the following two paragraphs. Then, shot boundary detection and the selection of a subsequence of frames are outlined. Finally, some results are presented and conclusions are drawn.

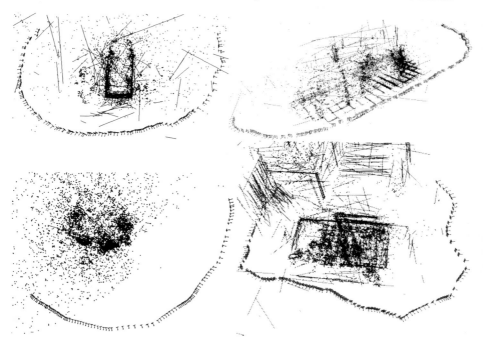

Fig. 1. Examples of large reconstructions. Top left: A birds perspective on a car (Volvo). Top right: Five bicycles (Bikes). Bottom left: Four girls standing in a half-circle (Girlsstatue). Bottom right: The author's breakfast table (Swedish Breakfast). Some frames from each sequence can be found in Figures 17-20

2 Global Motion Estimation

The global motion estimation is done using the initial step of a coarse to fine, optical flow based, video mosaicing algorithm [13,17,21,25]. The motivations behind using a flow based approach over a feature based (such as e.g. [3]) in this case were that the behaviour is good also for gravely unsharp frames and that it is easy to obtain fast approximations by downsampling. The motion model is an arbitrary rotation of the camera around the centre of projection and an arbitrary change of linear calibration. Assuming also a rigid world, this is equivalent to a homographic mapping H, represented by a 3x3 matrix, between the homogenous image coordinates x_1 and x_2 of the first and second frame as

$$x_2 \cong Hx_1 , \qquad (1)$$

where $\cong$ denotes equality up to scale. Both the images are downsampled to a small size of, for example, 50x50 pixels. To avoid problems due to altered lighting and overall brightness, both images are also normalised to have zero mean and unit standard deviation. The mapping H has eight degrees of freedom and should be minimally parameterized. As only small rotation is expected, this can be done safely by setting $H_{33} = 1$. The minimisation criterion applied to the estimation of H is the mean square residual. Better measures exist [13], but here the objective is only to obtain a rough estimation quickly. The mean square residual R between the image functions f_1 and f_2, using H for the correspondence is

$$R(f_1, f_2, H, \Theta) = \frac{1}{\#(\Theta)} \sum_{x \in \Theta} (f_2(Hx) - f_1(x))^2 \ . \qquad (2)$$

Here, Θ is all or a subset of the set Θ_A of pixels in the first image that are mapped into the second image and $\#(\Theta)$ is the number of elements of Θ. Larger sets than Θ_A are also possible if the image functions are defined beyond the true image domain by some extension scheme. In this case, Θ was chosen to be the whole image, except for a border of width d, which is a maximal expected disparity. The unit matrix is used as the initial estimate of H. Then, R is minimised by a non-linear least squares algorithm such as Levenberg-Marquardt [19].

3 Sharpness Measure

The measure of image sharpness is a mean square of the horizontal and vertical derivatives, evaluated as finite differences. More exactly

$$S(f, I) = \frac{1}{2 \#(I)} \sum_{(x,y) \in I} (f(x+1, y) - f(x-1, y))^2 + (f(x, y+1) - f(x, y-1))^2 \ , \qquad (3)$$

where I is the whole image domain except for the image boundaries. This measure is not used in any absolute sense, but only to measure the relative sharpness of similar images.

4 Shot Boundary Detection

The first step of the preprocessor is to detect shot boundaries. These occur when the camera has been stopped and then started again at a new position. This information could of course be provided from the camera, but in practice this is not always the case. It should be mentioned that since a video sequence is discretely sampled in time, a shot boundary is not strictly defined. With high frame rate material, the shot boundaries can be detected rather reliably. At lower frame rates however, the distinction between a shot boundary and a large camera motion becomes somewhat arbitrary. Shot boundaries are detected by evaluating the correlation between adjacent frames after global motion compensation. If the correlation is below a threshold, the second image is declared the beginning of a new shot. The correlation could be measured by the same mean square measure that was used for the motion estimation, but here it was preferred to use the normalised correlation coefficient, as this yields a more intuitively interpretable value. The threshold is currently set to $t_1 = 0.75$.

5 Selection of a Subsequence of Frames

Once the shot boundaries are detected, processing proceeds independently for each shot. While the shot boundary detection is performed in a purely sequential manner, the algorithm for selecting key frames operates in a batch mode. The algorithm was tested and will be described with each shot as one batch. The algorithm is conceptually simple and can almost be summarised in a single sentence:

Traverse all frames in order of increasing sharpness and delete redundant frames.

To avoid confusion, this is rephrased in algorithmic form. The letter Ω will be used to denote the subsequence of frames that remain at a particular time.

1. Set Ω to the sequence of all frames. Create a list L with the frames of Ω in order of increasing sharpness.
2. For all frames F_i in L do:

    If F_i is redundant in Ω, then remove F_i from Ω.

It remains to define when a frame is redundant in the subsequence Ω. A frame is redundant in Ω if it is not essential for the connectivity of Ω, as follows. Consider the frame F_i, belonging to the subsequence $\Omega = \{F_n\}_{n=1}^N$ of frames that remain at this particular time. If $i = 1$ or $i = N$, the frame is not redundant. Otherwise, global motion estimation is performed past the frame, i.e. between frame F_{i-1} and frame F_{i+1}. If this motion estimation yields a final correlation coefficient above a threshold, currently set to $t_2 = 0.95$ and the estimated mapping H does not violate the maximum expected disparity d at any point, the frame F_i is redundant. The value of d is currently set to ten percent of the image size, which is half of the maximum disparity expected by the SaM algorithm.

With the above scheme, frames are deleted until further deletions would cause too high discrepancies between neighbouring frames. Observe that frames that are considered early for deletion are more likely to become removed, since the subsequence Ω is then very dense. The traversal in order of increasing sharpness therefore ensures that the preprocessor prefers keeping sharp frames. The discrepancy that prevents a deletion can be either a violation of the disparity constraint or significant parallax that causes the global motion estimation, with the assumption of camera rotation, to break down. In the latter case, the material has become suitable for a SaM algorithm. In the former case, the material is ready for SaM or possibly mosaicing.

6 Results

Attention is first turned to a theoretical result. The preprocessing algorithm is approximately idempotent and can be made perfectly idempotent by a modification. Instead of only executing one run over all frames to perform deletions, this run is repeated until no additional deletions occur. The algorithm is now perfectly idempotent. To see why, consider application of the preprocessor a second time. No shot boundaries will be detected, because all adjacent frames with a correlation less than t_1 after motion compensation were detected during the first pass and no new such pairs have been created by frame deletion, since $t_2 > t_1$. Neither do any frame deletions occur during the second pass, since this was the stopping criterion for the first pass.

Let us now turn to the practical experiments. The results of a preprocessor are not strictly measurable unless the type of subsequent processing is defined. The experiments were performed in conjunction with a feature based SaM algorithm, similar in spirit to, for instance [1,2,8,10,18,27]. Details can be found in [16]. The algorithm takes a video sequence and automatically extracts a sparse representation in terms of points and lines of the observed structure. It also estimates camera position, rotation and calibration for all frames. The preprocessor was tested on approximately 50 sequences, most of them handheld with jerky motion and imperfect focus. In this paper, results from the sequences listed in Table 1 have been or will be cited. Some frames from the sequences are shown in Figures 10-21.

As was mentioned in the introduction, the preprocessor should not significantly change data that does not need preprocessing. This was tested in practice by applying the preprocessor and subsequent SaM system to sequences with sharp, nicely separated frames and no shot boundaries. Final reconstruction results for the sequences House and Basement are shown in Figure 2. For the House sequence, the preprocessor does not falsely detect any shot boundaries, nor does it remove any frames. In other words, it just propagates the input data to its output, which is exactly the desired behaviour. In the final reconstruction, three camera views are missing at the end of the camera trajectory, but these views are removed by the SaM algorithm and not by the preprocessor. The textures shown in this paper are created with a very tentative algorithm using only one of the camera views. The textures are included to facilitate interpretation of the reconstructions. A dense reconstruction scheme is under development.

Table 1. Data for test sequences

Name	Length	Resolution	Type
House	10	768 x 576	Turntable
Basement	11	512 x 512	Autonom. Vehicle
Sceneswap	748	352 x 288	Handheld
TriScene	315	"	"
Room	99	"	"
Stove	107	"	"
David	19	"	"
Volvo	525	"	"
Bikes	161	"	"
Girlsstatue	541	"	"
Swedish Breakfast	363	"	"
Nissan Micra	340	"	"

Fig. 2. Final reconstructions from the sequences House and Basement

The preprocessor did not falsely detect any shot boundaries in the sequence Basement either. However, it deleted frames 3 and 7, which can in fact be seen as larger gaps in the camera trajectory. This happens because the forward motion does not cause enough parallax. It does not negatively affect the final result.

Experimental results of shot boundary detection on the sequence Sceneswap is shown in Figure 3. This sequence consists of eleven shots, separated by shot boundaries after frame 72, 164, 223, 349, 423, 465, 519, 583, 619 and 681 (found manually). The threshold at 0.75 is shown as a solid line. Results are given at frame rates 25, 6.25 and 3.125 Hz. At all frame rates, the ten boundaries are found successfully and can be seen as ten groups of three markers below the detection threshold at the above mentioned frame numbers. At 25 and 6.25 Hz the detection is stable, with a correlation above 0.95 and 0.9, respectively, for all non-boundaries. This can be seen as a pattern at the top of the figure. At 3.125 Hz however, the frame rate has dropped too low and five false responses occur, all marked with an arrowhead. Thus the importance of a high input frame rate is illustrated.

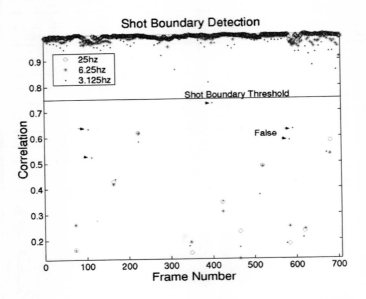

Fig. 3. Result of shot boundary detection on the sequence Sceneswap

A typical preprocessor result is shown in Figure 4 for the 25 Hz sequence TriScene, with two correctly detected shot boundaries. The frames surviving the decimation are marked by triangles. Sharpness is on the vertical axis. Observe that local sharpness minima are avoided.

In Figure 5, it is illustrated how the preprocessor manages to make the system independent of the input frame rate, provided that this is sufficiently high. The result is for the 12.5 Hz sequence Stove, with a total of 107 frames. The sequence is handheld, with the camera moving in an arc in front of a kitchen stove. The sequence was subsampled to 50 lower frame rates and fed into the preprocessor. With very few input frames (<12), shot boundaries are falsely detected. With the number of input frames higher than 30 however, this is no longer a problem and the number of output frames remains fairly constant at about 20. When fed with the full frame rate, the preprocessor removes about 80% of the frames and the SaM algorithm can then carry on to produce the reconstruction shown in Figure 6.

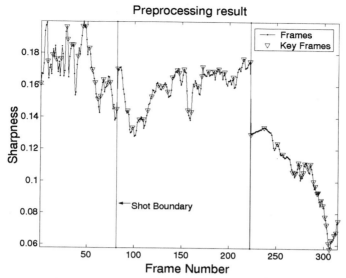

Fig. 4. Preprocessor result for the sequence TriScene

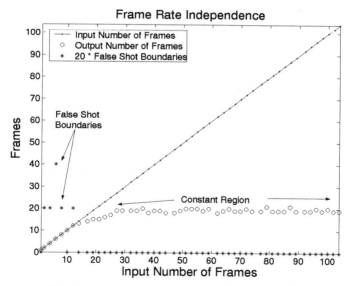

Fig. 5. Frame rate independence for the sequence Stove

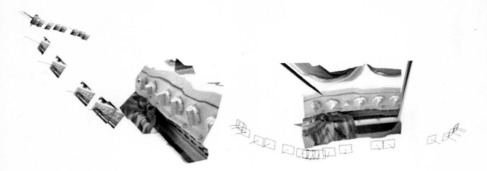

Fig. 6. Final reconstruction of Stove

In order to characterise the behaviour of the frame decimation algorithm for various amounts of decimation, the frame decimation was performed on the complete video material with a number of choices for the redundancy correlation threshold t_2 and the maximum expected disparity d. The result for the sequence Stove is shown in Table 2. The left column shows the decimation thresholds with d in parts of the image size. The column 'Frames' shows the number of frames that are left after decimation with these thresholds. The columns 'Time' and 'Mem' indicate the running time and the maximum memory usage of the SaM algorithm. The code is not optimised for speed or low memory usage, but the data still gives an indication of how the problem grows. The number of points and lines in the reconstruction are displayed in the columns 'Points' and 'Lines'. 'P_error' is the root mean square point reprojection error in number of pixels. 'L_error' is the root mean square line reprojection error. The line reprojection error is measured as the length of the vector $l - \hat{l}$, where l and $\hat{l}$ are the observed and reprojected line, represented as homogenous line vectors normalised to hit the unit cube. Observe that some results are marked with a *. This means that the SaM algorithm did not manage to build a complete Euclidian reconstruction of the decimated sequence. With decimation down to only 9 frames, the connectivity of the sequence is lost. With very little decimation, the problem is very large and many unfocused frames are still included. Therefore, at 103 frames the reconstruction fails for the second part of the sequence and at 107 frames the reconstruction fails completely. In figure 7 the reconstructions corresponding to all rows of the table, except the first and last row, are shown visually. Note that the camera trajectory displays the same characteristics in all cases except the last, where the reconstruction failed.

Table 2. Variable amount of frame decimation on the sequence Stove. The table is explained above.

$t_2;d$	Frames	Time(s)	Mem	Points	Lines	P_error	L_error
0.4;0.35	9	3986*	15M*	146*	3*	0.771*	0.071*
0.7;0.3	12	4340	15M	469	26	0.652	0.016
0.8;0.15	16	6580	16M	823	41	0.672	0.018
0.9;0.125	18	11827	16M	988	70	0.688	0.019
0.95;0.1	20	11016	16M	1017	63	0.674	0.021
0.975;0.075	25	18647	17M	1301	92	0.722	0.022
0.982;0.062	32	12269	17M	1634	97	0.762	0.022
0.99;0.05	61	31623	26M	2632	166	0.809	0.020
0.992;0.042	76	88123	32M	3047	196	0.820	0.013
0.9935;0.04	91	123120	35M	3466	210	0.831	0.015
0.995;0.037	103	93255*	31M*	1315*	66*	0.597*	0.026*
0.999;0.025	107	103874*	51M*	*	*	*	*

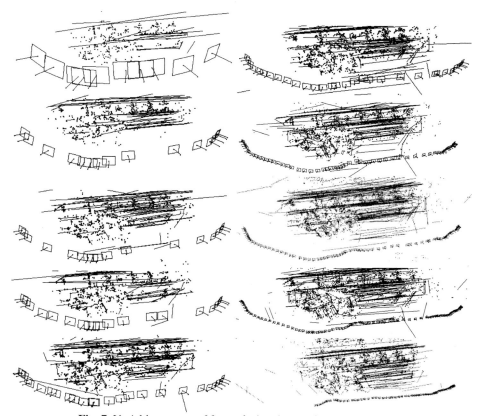

Fig. 7. Variable amount of frame decimation on the sequence Stove

In Figure 8, the reconstruction from the sequence Room is shown. This is a handheld sequence, where the camera moves forward through an office. Many frames are out of focus. At the beginning and the end of the sequence, the camera moves very little and only rotates slightly back and forth, which is not uncommon in raw video material. The preprocessor successfully removes most of these frames, which enables a reasonable trajectory of camera views to be extracted, although the structure for this sequence is still very poor.

Fig. 8. Final reconstruction of Room

In Figure 9 reconstructions from the sequences Nissan Micra and David are shown. The sequence David was acquired by holding the camera with a stretched arm and performing an arched motion. Again, the motion of the centre of projection is negligible between a couple of frames near the end of the arc. Depending on the SaM system, this can sometimes cause problems with degeneracy. With frame decimation, the troublesome frames are removed.

7   Conclusions

A preprocessor that performs shot boundary detection followed by frame decimation has been proposed. The results show that using this preprocessor in a SaM system has several benefits. By an automatic decimation of the number of frames used for the actual SaM calculations, it is possible to keep the size of the problem manageable, independently of the input frame rate. This makes it feasible to use a high input frame rate, which in turn avoids loss of connectivity due to matching difficulties. The high input frame rate also enables robust detection of shot boundaries. Indications have been given that the proposed type of preprocessor sometimes can eliminate problems of degeneracy or near degeneracy due to insufficient motion. It has been discussed why the preprocessor algorithm is approximately idempotent and how it can be made exactly idempotent by a modification. It was also shown that the preprocessor does not have a negative impact on material that already represents good input to a SaM algorithm. In summary, the proposed frame decimation makes automatic extraction of SaM from handheld video sequences more practical.

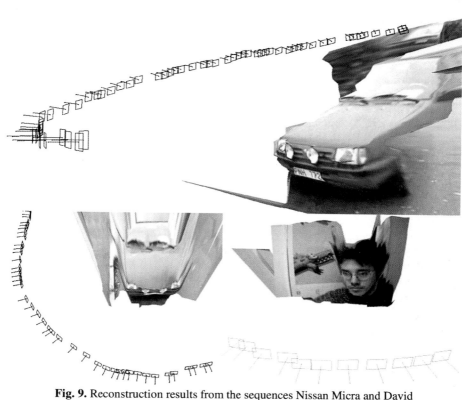

Fig. 9. Reconstruction results from the sequences Nissan Micra and David

Acknowledgements. The author is grateful to Professor Andrew Zisserman of Oxford Visual Geometry Group for making the House and Basement sequences available. Thanks also to Rickard Sjöberg for help with the handheld sequences.

References

1. P. Beardsley, P. Torr, A. Zisserman, 3D model acquisition from extended image sequences, *Proc. ECCV 96*, pp. 683-695.
2. P. Beardsley, A. Zisserman, D. Murray, Sequential updating of projective and affine structure from motion, *IJCV*, 23(3), pp. 235-259, 1997.
3. D. Capel, A. Zisserman, Automated mosaicing with super-resolution zoom, *Proc. CVPR 98*, pp. 885-891.
4. R. Cipolla, E. G. Boyer, 3D model acquisition from uncalibrated images, *Proc IAPR Workshop on Machine Vision Applications*, Chiba Japan, pp. 559-568, November 1998.
5. P. Debevec, C. Taylor, J. Malik, Modeling and rendering architecture from photographs: A hybrid geometry- and image-based approach, *Proc. SIGGRAPH'96*, pp. 11-20.
6. O. Faugeras, What can be seen in three dimensions with an uncalibrated stereo rig?, *Proc. ECCV 92*, pp. 563-578.

7. P. Fua, Reconstructing complex surfaces from multiple stereo views, *Proc. ICCV 95*, pp. 1078-1085.
8. A. W. Fitzgibbon, A. Zisserman, Automatic camera recovery for closed or open image sequences, *Proc. ECCV 98*, pp. 311-326.
9. K. Hanna, N. Okamoto, Combining stereo and motion for direct estimation of scene structure, *Proc. ICCV 93*, pp. 357-365.
10. R. Hartley, Euclidean reconstruction from uncalibrated views, *Applications of Invariance in Computer Vision*, LNCS 825, pp. 237-256, Springer-Verlag, 1994.
11. A. Heyden, K. Åström, Euclidean reconstruction from image sequences with varying and unknown focal length and principal point, *Proc. CVPR 97*, pp. 438-443.
12. M. Irani, P. Anandan, M. Cohen, Direct recovery of planar-parallax from multiple frames, *Proc. Vision Algorithms Workshop (ICCV 99)*, Corfu Greece, pp. 1-8, September 1999.
13. K. Kanatani, N. Ohta, Accuracy bounds and optimal computation of homography for image mosaicing applications, *Proc. ICCV 99*, pp. 73-78.
14. P. McLauchlan, D. Murray, A unifying framework for structure from motion recovery from image sequences, *Proc. ICCV 95*, pp. 314-320.
15. R. Mohr, F. Veillon, L. Quan, Relative 3D reconstruction using multiple uncalibrated images, *Proc. CVPR 93*, pp. 543-548.
16. D. Nistér, Reconstruction from uncalibrated sequences with a hierarchy of trifocal tensors, *Accepted to ECCV 2000*.
17. S. Peleg, J. Herman, Panoramic mosaics by manifold projection, *Proc. CVPR 97*, pp. 338-343.
18. M. Pollefeys, R. Koch, L. Van Gool, Self-calibration and metric reconstruction in spite of varying and unknown internal camera parameters, *IJCV*, 32(1), pp. 7-26, Aug, 1999.
19. W. Press, S. Teukolsky, W. Vetterling, B. Flannery, *Numerical recipes in C*, ISBN 0-521-43108-5, Cambridge University Press, 1988.
20. L. Robert, O. Faugeras, Relative 3D positioning and 3D convex hull computation from a weakly calibrated stereo pair, *Proc. ICCV 93*, pp. 540-544.
21. H. Sawhney, S. Hsu, R. Kumar, "Robust video mosaicing through topology inference and local to global alignment", *Proc. ECCV 98*, pp.103-119.
22. A. Shashua, Trilinearity in visual recognition by alignment, *Proc. ECCV 94*, pp. 479-484.
23. M. Spetsakis, J. Aloimonos, Structure from motion using line correspondences, *IJCV*, pp. 171-183, 1990.
24. P. Sturm, W. Triggs, A factorization based algorithm for multi-image projective structure and motion, *Proc. ECCV 96*, pp. 709-720.
25. R. Szeliski, H.-Y. Shum, Creating full view panoramic image mosaics and environment maps, *Proc. SIGGRAPH'97*, pp. 251-258.
26. C. Tomasi, T. Kanade, Shape and motion from image streams under orthography: a factorization approach, *IJCV*, 9(2), pp. 137-154. November 1992.
27. L. Van Gool, A. Zisserman, Automatic 3D model building from video sequences, *Proc. ECMAST 96*, pp. 563-582.

Fig. 10. Some frames from House

Fig. 11. Some frames from Basement

Fig. 12. One frame from each shot of the sequence Sceneswap

Fig. 13. Two frames from each shot of the sequence TriScene

Fig. 14. Some frames from Room

Fig. 15. Some frames from Stove

Fig. 16. Some frames from David

Fig. 17. Some frames from Volvo

Fig. 18. Some frames from Bikes

Fig. 19. Some frames from Girlsstatue

Fig. 20. Some frames from Swedish Breakfast

Fig. 21. Some frames from Nissan Micra

Discussion

1. **Hans-Helmut Nagel, Universität Karlsruhe:** I wonder to what extent you could use techniques in video compression to detect shot boundaries. They have the same problem if they do high compression: they want to detect shot boundaries in order to set-up their system anew. So, could you use these techniques and, if not, I would be interested to learn the reasons.
David Nister: Do you mean the work that has been done on shot detection and reference view selection in for example MPEG-related activities? Well, certainly there has been a lot of work done on that and the motive is usually to segment and summarize the material. For example you want to send just a few frames of a news sequence to a mobile terminal. I think that the shot detection techniques

is not tuned to structure and motion. It gives too disparate views and it is not concerned about matching. I want output that can be subsequently matched and I tend to keep many more frames than is done in that context. There is also work done on reference view selection for view synthesis, but it is then usually assumed that the cameras are calibrated extrinsically and intrinsically beforehand and I do not want to do this since my main motive is to limit the computational complexity and as a consequence, view selection is the first thing I do.

2. **Rick Szeliski, Microsoft:** I like the idea that you have of not keeping the motion blurred frames. That seems like a good idea although for feature tracking it may not be that important. It is a nice framework. The one thing I am a little puzzled by is that you tend to keep more frames when the camera motion is large and most of that motion, which you call jerkiness, is due to pure rotation. So, if you are already computing the homographies, why not map the images through the homographies before running your feature tracker. I realize that feature trackers break down when the motion is large, but if you just warp the two images that you are tracking by the homography it is only the total amount of parallax (in other words, large translational motion) were you need dense sampling and the rotational motion almost irrelevant, just as Bill Triggs showed [1]. You just want to get rid of the rotation. You want to stabilize the sequence. So, why not stabilize before running the rest of your algorithm?

David Nister: It is a good point that the homographies that are estimated could be used to stabilize by removing undesired rotational motion. My motivation for not doing this is the following. I use the homography motion model to quickly verify that some frames are redundant so that I can dispose of them. However, I want to be able to handle all types of sequences, including ones where there is a large amount of parallax between consecutive frames. The homography model does not fit well to that type of sequence. The accuracy of the homographies might therefore be impaired. As I do things now, this will result in most frames being forwarded to the structure from motion system, which can handle the parallax. This is the desired behavior and will not cause any problems. If, on the other hand, the inaccurate homographies were used for stabilization, it might cause more problems than it solves.

3. **Paul Debevec, University of Southern California:** I was wondering if on your video camera you could not set the shutter to thousandths of a second so that you do not get motion blurring. Does that not work because you still get the interlacing with the two fields not matching up?

David Nister: I guess that changing the shutter speed will definitely help if only one field is used. However, I believe that blurring is inevitable in the type of amateur material that I want to be able to handle. I also think the blurring in my sequences is not always motion blur. It is rather common with jerky camera motion that the auto-focus loses track of things and it then takes a while before it finds its way.

4. **Tomas Pajdla, Czech Technical University:** Is it a good idea to make your selection more dependent on the amount of occlusion in the scene or is this somehow implicitly taken into account by correlation, which you use? Because if

you have more occlusion, you probably need more frames to get through the structure.

David Nister: The frame decimation is the first processing I do on a sequence, so I do not know the structure or the occlusions.

Tomas Pajdla: Yes, I know that you do not know it, but you estimate it. If you have more occlusion, more complex structure, you will probably have to use more frames.

David Nister: That is right, and this is taken into account in the sense that if there is a lot of occlusion, the homography is not enough to compensate between frames, leading to a low correlation and thus also more frames. The selection would of course benefit from more precise knowledge of the structure, but this is estimated much higher up in the system. The frame decimation requires on the order of a second per frame, while the structure and motion system uses on the order of minutes per frame, so I do not know the exact occlusions until much later.

References

1. B. Triggs. Plane + Parallax, Tensors and Factorization. In *Proc. European Conference on Computer Vision*, pages 522-538, 2000.

Stabilizing Image Mosaicing by Model Selection

Yasushi Kanazawa[1] and Kenichi Kanatani[2]

[1] Department of Knowledge-based Information Engineering
Toyohashi University of Technology
1-1 Hibarigaoka, Tenpaku, Toyohashi, Aichi 441-8580 Japan
kanazawa@tutkie.tut.ac.jp

[2] Department of Computer Science, Gunma University
1-5-1, Tenjin, Kiryu, Gunma 376-8515 Japan
kanatani@cs.gunma-u.ac.jp

Abstract. The computation for image mosaicing using homographies is numerically unstable and causes large image distortions if the matching points are small in number and concentrated in a small region in each image. This instability stems from the fact that actual transformations of images are usually in a small subgroup of the group of homographies. It is shown that such undesirable distortions can be removed by *model selection* using the geometric AIC without introducing any empirical thresholds. It is shown that the accuracy of image mosaicing can be improved *beyond* the theoretical bound imposed on statistical optimization. This is made possible by our *knowledge* about probable subgroups of the group of homographies. We demonstrate the effectiveness of our method by real image examples.

1 Introduction

Image mosaicing is a technique for integrating multiple images into one continuous image, a typical one being a *panoramic image* [10,13,15,16]. This technique has long been used for creating terrain maps from aerial images or analyzing remote sensing satellite images, but recently its applications to virtual reality creation from multiple scene images are attracting much attention. Image mosaicing also plays an important role in automatic surveillance using camera images.

The basic principle underlying image mosaicing is the computation of a *homography*, which is a mapping that typically occurs between two perspective images of a planar surface in the scene [3]. Since faraway scenes can effectively be regarded as planar surfaces, we can register one image to another by computing the homography between them.

If the images have very small overlaps between them, as is often the case for remote sensing images and aerial images, only a small number of matching points are available. In such a case, the selected points in one image may be mapped to the corresponding points in the other image fairly accurately, but if we extrapolate this mapping to portions apart from the matching points, a large distortion may occur even in the presence of very small noise (Fig. 1(a)). Since a homography may map some points to infinity, the part beyond those points can appear from the other side of the image frame (Fig. 1(b)).

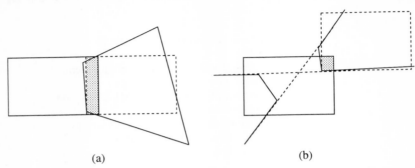

Fig. 1. (a) If images with a very small overlap are used for mosaicing, a large distortion may result even in the presence of small noise. (b) Some part of the image may appear from the other side of the frame.

In [7] this instability was demonstrated by using real images. An accurate algorithm was also presented for computing a homography from point correspondences using a technique called *renormalization*, which not only produces a statistically optimal solution but also evaluates the reliability of the computed solution in quantitative terms. The algorithm is implemented in C++ and publicly available via the Web[1]. A theoretical accuracy bound was also derived for the homography computation. It was experimentally confirmed that the renormalization algorithm indeed produces estimates in the vicinity of that bound.

Although the renormalization algorithm dramatically reduces the instability of the mapping, as demonstrated in [7], it cannot remove the distortion completely. However, further improvement is theoretically impossible with this technique. In this paper, we will show that this limitation *can* be broken through by incorporating our *knowledge* about the source of the instability. The instability stems from the fact that while homographies constitute an 8-parameter group of transformations, actual transformations are usually in a small subgroup, e.g., the group of translations, the group of rigid motions, the group of similarities, or the group of affine transformations. In the presence of noise, the computed solution moves out of the subgroup to which it should belong, causing a large image distortion.

In the following, we show that such undesirable distortions can be removed by *model selection* using the *geometric AIC* [4,5] without introducing any empirical thresholds. We also present a Levenberg-Marquardt scheme for optimization and an analytical procedure for computing an initial guess. We demonstrate the effectiveness of our method by real image examples.

---

[1] http://www.ail.cs.gunma-u.ac.jp/Labo/programs-e.html

2 Representation of Homography

A *homography* is an image mapping expressed in the following form:

$$x' = \frac{Ax + By + C}{Px + Qy + R}, \qquad y' = \frac{Dx + Ey + F}{Px + Qy + R}. \tag{1}$$

If we define vectors x and x' and matrix H by

$$\boldsymbol{x} = \begin{pmatrix} x/f_0 \\ y/f_0 \\ 1 \end{pmatrix}, \quad \boldsymbol{x}' = \begin{pmatrix} x'/f_0 \\ y'/f_0 \\ 1 \end{pmatrix}, \quad \boldsymbol{H} = \begin{pmatrix} A & B & C/f_0 \\ D & E & F/f_0 \\ P/f_0 & Q/f_0 & R/f_0^2 \end{pmatrix}, \tag{2}$$

eq. (1) can be written as

$$\boldsymbol{x}' = Z[\boldsymbol{H}\boldsymbol{x}]. \tag{3}$$

Here, $Z[\,\cdot\,]$ denotes normalization to make the third element 1; f_0 is a scale factor chosen so that x/f_0 and y/f_0 have order 1.

Given two images, we choose matching points from them. Let $\{(x_\alpha, y_\alpha)\}$ and $\{(x'_\alpha, y'_\alpha)\}$ be the coordinates of the points chosen from the first and the second images, respectively. Let $\{\boldsymbol{x}_\alpha\}$ and $\{\boldsymbol{x}'_\alpha\}$ be their vector representations. We regard them as random Gaussian variables with covariance matrices $V[\boldsymbol{x}_\alpha]$ and $V[\boldsymbol{x}'_\alpha]$.

The absolute magnitude of noise is difficult to predict a priori, but its geometric characteristics such as homogeneity/inhomogeneity and isotropy/anisotropy can be relatively easily predicted. For example, if we use template matching for finding corresponding points, the uncertainty of matching is measured by the Hessian of the residual surface around the detected point [2,9,11,12]. Here, we assume that the covariance matrices $V[\boldsymbol{x}_\alpha]$ and $V[\boldsymbol{x}'_\alpha]$ are known *up to scale* and write

$$V[\boldsymbol{x}_\alpha] = \epsilon^2 V_0[\boldsymbol{x}_\alpha], \qquad V[\boldsymbol{x}'_\alpha] = \epsilon^2 V_0[\boldsymbol{x}'_\alpha]. \tag{4}$$

We call the unknown magnitude ϵ the *noise level*. The matrices $V_0[\boldsymbol{x}_\alpha]$ and $V_0[\boldsymbol{x}'_\alpha]$, which we call the *normalized covariance matrices*, specify the relative dependence of noise occurrence on positions and orientations. If no a priori knowledge is available for them, we simply assume isotropy and homogeneity and input the default values $V_0[\boldsymbol{x}_\alpha] = V_0[\boldsymbol{x}'_\alpha] = \text{diag}(1, 1, 0)$ (the diagonal matrix whose diagonal elements are 1, 1, and 0 in that order).

3 Optimal Homography Estimation

Eq. (3) can equivalently be written in the form $\boldsymbol{x}' \times \boldsymbol{H}\boldsymbol{x} = \boldsymbol{0}$. Hence, the task is to estimate $\boldsymbol{H}$ from noisy data $\{\boldsymbol{x}_\alpha\}$ and $\{\boldsymbol{x}'_\alpha\}$ with the knowledge that their true values $\{\bar{\boldsymbol{x}}_\alpha\}$ and $\{\bar{\boldsymbol{x}}'_\alpha\}$ satisfy

$$\bar{\boldsymbol{x}}'_\alpha \times \boldsymbol{H}\bar{\boldsymbol{x}}_\alpha = \boldsymbol{0}. \tag{5}$$

The reliability of an estimate $\hat{\boldsymbol{H}}$ of $\boldsymbol{H}$ can be measured by its *covariance tensor* $\mathcal{V}[\hat{\boldsymbol{H}}]$. A theoretical lower bound on it can be derived in analytical terms [7].

It is well known [4] that an optimal estimate of H which attains the accuracy bound in the first order (i.e., if terms of $O(\epsilon^4)$ are ignored) can be obtained by *maximum likelihood estimation*, minimizing the squared *Mahalanobis distances*

$$J = \frac{1}{N}\sum_{\alpha=1}^{N}(\boldsymbol{x}_\alpha - \bar{\boldsymbol{x}}_\alpha, V_0[\boldsymbol{x}_\alpha]_2^-(\boldsymbol{x}_\alpha - \bar{\boldsymbol{x}}_\alpha)) + \frac{1}{N}\sum_{\alpha=1}^{N}(\boldsymbol{x}'_\alpha - \bar{\boldsymbol{x}}'_\alpha, V_0[\boldsymbol{x}'_\alpha]_2^-(\boldsymbol{x}'_\alpha - \bar{\boldsymbol{x}}'_\alpha)), \quad (6)$$

subject to the constraint (5). Here and throughout this paper, $(\boldsymbol{a}, \boldsymbol{b})$ denotes the inner product of vectors $\boldsymbol{a}$ and $\boldsymbol{b}$. The super script $(\cdot)_r^-$ denotes the (Moore-Penrose) generalized inverse computed after replacing the smallest $n-r$ eigenvalues by zeros.

Using Lagrange multipliers and introducing first order approximation, we can eliminate the constraint (5) and express eq. (6) in the following form [4]:

$$J = \frac{1}{N}\sum_{\alpha=1}^{N}(\boldsymbol{x}'_\alpha \times H\boldsymbol{x}_\alpha, W_\alpha(\boldsymbol{x}'_\alpha \times H\boldsymbol{x}_\alpha)), \quad (7)$$

$$W_\alpha = \left(\boldsymbol{x}'_\alpha \times HV_0[\boldsymbol{x}_\alpha]H^\top \times \boldsymbol{x}'_\alpha + (H\boldsymbol{x}_\alpha) \times V_0[\boldsymbol{x}'_\alpha] \times (H\boldsymbol{x}_\alpha)\right)_2^-. \quad (8)$$

Let $\hat{J}$ be the residual, i.e., the minimum of the function J. It can be shown that $\hat{J}/\epsilon^2$ is subject to a χ^2 distribution with $2(N-4)$ degrees of freedom to a first approximation [4]. Hence, an unbiased estimator of ϵ^2 is obtained in the form

$$\hat{\epsilon}^2 = \frac{\hat{J}}{2(1-4/N)}. \quad (9)$$

In [7] a computational technique called *renormalization* was presented. It was experimentally confirmed that the solution practically falls on the theoretical accuracy bound.

4 Models of Image Transformations

Since the elements of H have scale indeterminacy (see eq. (3)), a homography has eight degrees of freedom. However, image transformations that we often encounter have much smaller degrees of freedom. For example, if a moving camera takes images of a faraway scene with varying zooming, the translation of the camera causes no visible changes, so the image transformation is parameterized by the camera rotation R (three degrees of freedom) and the focal lengths f and f' of the two frames (Fig. 2). Such transformations constitute a 5-parameter subgroup of the 8-parameter group of homographies. If the focal length is fixed, we obtain a 4-parameter subgroup.

If the camera translates relative to a nearby scene, we have the group of translations with two degrees of freedom (Fig. 3(b)). If the camera is allowed to rotate around its optical axis, we have the group of 2-D rigid motions with three degrees of freedom (Fig. 3(c)). If the focal length is also allowed to change, we have the group of similarities with four degrees of freedom (Fig. 3(d)). If the object is a planar surface in the distance, the image transformation can be viewed as an affine transformation with six degrees of

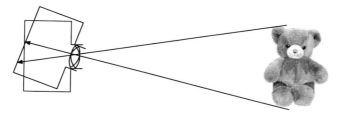

Fig. 2. Image transformation due to camera rotation.

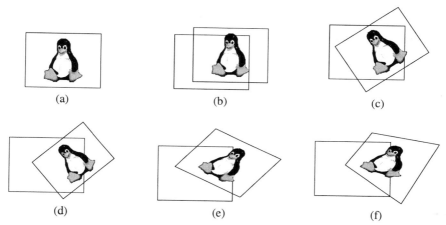

Fig. 3. (a) Original image. (b) Translation. (c) Rigid motion. (d) Similarity. (e) Affine transformation. (f) Homography.

freedom (Fig. 3(e)). All these image transformations belong to a subgroup of the group of general homographies (Fig. 3(f)).

Thus, we have a hierarchy of image transformations (Fig. 4). In the presence of noise, however, the computed homography need not necessarily belong to the required subgroup, resulting in a large image distortion that cannot be attributed to any camera motion. Such a distortion can be removed if we find a homography within the required subgroup or *model*. For example, if the image transformation is known to be a 2-D rigid motion, we only need to compute the image rotation and translation optimally. However, we do not know *a priori* to which model the observed transformation should belong.

A naive idea is to choose from among candidate models the one that gives the smallest residual. This does not work, however, because the 8-parameter homography group is always chosen: a model with more degrees of freedom has a smaller residual. For a fair comparison, we need to compensate for the overfit caused by excessive degrees of freedom. Here, we measure the goodness of a model by the *geometric AIC* [4,5,6], which is a special form of Akaike's *AIC* [1]. The model with the smallest geometric AIC is preferred. See [14] for other model selection criteria.

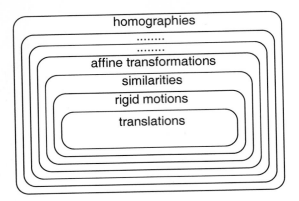

Fig. 4. Hierarchy of image transformations.

5 Subgroup Hierarchy

5.1 8-Parameter Homographies

Let $\hat{J}_{H_8}$ be the resulting residual of eq. (7). The geometric AIC is given by

$$\text{G-AIC}_{H_8} = \hat{J}_{H_8} + \frac{16}{N}\epsilon^2, \tag{10}$$

where the square noise level ϵ^2 is estimated by eq. (9).

5.2 5-Parameter Homographies

If the camera rotates by R around the center of the lens and changes the focal length from f to f', the resulting homography has the form

$$H = F'^{-1} R^\top F, \tag{11}$$

where

$$F = \text{diag}(1, 1, \frac{f}{f_0}), \qquad F' = \text{diag}(1, 1, \frac{f'}{f_0}). \tag{12}$$

We use the Levenberg-Marquardt method (LM method) for minimizing eq. (7). First, we define the following non-dimensional variables:

$$\phi = \frac{f}{f_0}, \qquad \phi' = \frac{f'}{f_0}. \tag{13}$$

The minimization procedure goes as follows:

1. Let c=0.001. Analytically compute initial guesses of ϕ, ϕ', and R (see Appendix A), and evaluate the residual $J = J(\phi, \phi', R)$.

2. Compute the gradient ∇J and Hessian $\nabla^2 J$ (see Appendix B for the detailed expressions):

$$\nabla J = \begin{pmatrix} \partial J/\partial \phi \\ \partial J/\partial \phi' \\ \nabla_\Omega J \end{pmatrix}, \qquad (14)$$

$$\nabla^2 J = \begin{pmatrix} \partial^2 J/\partial \phi^2 & \partial^2 J/\partial \phi \partial \phi' & (\nabla_\Omega \partial J/\partial \phi)^\top \\ \partial^2 J/\partial \phi' \partial \phi & \partial^2 J/\partial \phi'^2 & (\nabla_\Omega \partial J/\partial \phi')^\top \\ \nabla_\Omega \partial J/\partial \phi & \nabla_\Omega \partial J/\partial \phi' & \nabla^2_\Omega J \end{pmatrix}. \qquad (15)$$

3. Let D be the diagonal matrix consisting of the diagonal elements of $\nabla^2 J$, and solve the following simultaneous linear equation:

$$(\nabla^2 J + cD) \begin{pmatrix} \Delta \phi \\ \Delta \phi' \\ \Delta \Omega \end{pmatrix} = -\nabla J. \qquad (16)$$

4. Compute the residual $J' = J(\phi + \Delta\phi, \phi' + \Delta\phi', \mathcal{R}(\Delta\Omega)R)$.
   - If $J > J'$, let $c \leftarrow 10c$ and go back to Step 3.
   - If $J < J'$ and $|J - J'|/J < \epsilon_J$, return ϕ, ϕ', and R and stop.
   - Else, let $c \leftarrow c/10$, update ϕ, ϕ', and R in the form

$$\phi \leftarrow \phi + \Delta\phi, \qquad \phi' \leftarrow \phi' + \Delta\phi', \qquad R \leftarrow \mathcal{R}(\Delta\Omega)R, \qquad (17)$$

and go back to Step 2.

Here, ϵ_J is a threshold for convergence, and $\mathcal{R}(\Delta\Omega)$ denotes the rotation around $\Delta\Omega$ by an angle $\|\Delta\Omega\|$. Let $\hat{J}_{H_5}$ be the resulting residual. The geometric AIC is given by

$$\text{G-AIC}_{H_5} = \hat{J}_{H_5} + \frac{10}{N}\epsilon^2, \qquad (18)$$

where the square noise level ϵ^2 is estimated by eq. (9).

5.3 4-Parameter Homographies

If we let $f = f'$ in eq. (11), we obtain the 4-parameter group of homographies, for which optimal values of ϕ and R are obtained by slightly modifying the LM method described above. Let $\hat{J}_{H_4}$ be the resulting residual. The geometric AIC is given by

$$\text{G-AIC}_{H_4} = \hat{J}_{H_4} + \frac{8}{N}\epsilon^2. \qquad (19)$$

5.4 Similarities

A similarity is a special homography that has the following form:

$$H = \begin{pmatrix} s\cos\theta & -s\sin\theta & t_1/f_0 \\ s\sin\theta & s\cos\theta & t_2/f_0 \\ 0 & 0 & 1 \end{pmatrix}. \qquad (20)$$

By this transformation, the image is rotated by angle θ around the origin, scaled by s, and translated by (t_1, t_2). If we define

$$\vec{x}_\alpha = \begin{pmatrix} x_\alpha/f_0 \\ y_\alpha/f_0 \end{pmatrix}, \qquad \vec{x}'_\alpha = \begin{pmatrix} x'_\alpha/f_0 \\ y'_\alpha/f_0 \end{pmatrix}, \tag{21}$$

$$R = \begin{pmatrix} \cos\theta & -\sin\theta \\ \sin\theta & \cos\theta \end{pmatrix}, \qquad \vec{\tau} = \begin{pmatrix} t_1/f_0 \\ t_2/f_0 \end{pmatrix}, \tag{22}$$

eq. (7) is rewritten in the following form:

$$J = \frac{1}{N} \sum_{\alpha=1}^{N} (\vec{x}'_\alpha - sR\vec{x}_\alpha - \vec{\tau}, W_\alpha(\vec{x}'_\alpha - sR\vec{x}_\alpha - \vec{\tau})), \tag{23}$$

$$W_\alpha = \left(s^2 R V_0[\vec{x}_\alpha] R^\top + V_0[\vec{x}'_\alpha] \right)^{-1}. \tag{24}$$

This is minimized by the LM method (we omit the details). See Appendix C for the procedure for computing an initial guess. Let $\hat{J}_S$ be the resulting residual. The geometric AIC is given by

$$\text{G-AIC}_S = \hat{J}_S + \frac{8}{N}\epsilon^2. \tag{25}$$

5.5 Rigid Motions

The image transformation reduces to a 2-D rigid motion if we let $s = 1$ in eq. (20). We can apply the same LM method for minimizing J and the procedure for computing an initial guess after an appropriate modification. Let $\hat{J}_M$ be the resulting residual. The geometric AIC is given by

$$\text{G-AIC}_M = \hat{J}_M + \frac{6}{N}\epsilon^2. \tag{26}$$

5.6 Translations

A 2-D rigid motion reduces to a translation if we let $\theta = 0$. Let $\hat{J}_T$ be the resulting residual. The geometric AIC is given by

$$\text{G-AIC}_T = \hat{J}_T + \frac{4}{N}\epsilon^2. \tag{27}$$

5.7 Affine Transformations

An affine transformation is a special homography that has the form

$$H = \begin{pmatrix} a_{11} & a_{12} & t_1/f_0 \\ a_{21} & a_{22} & t_2/f_0 \\ 0 & 0 & 1 \end{pmatrix}. \tag{28}$$

Optimal values of $\{a_{ij}\}$ and $\{t_i\}$ are obtained by the LM method, and an initial guess can be computed analytically (we omit the details). Let $\hat{J}_A$ be the resulting residual. The geometric AIC is given by

$$\text{G-AIC}_A = \hat{J}_A + \frac{12}{N}\epsilon^2. \tag{29}$$

Fig. 5. Real images of an outdoor scene and the selected points.

Fig. 6. (a) The image mapping computed by an optimal homography. (b) The image mapping by model selection.

5.8 Principle of Model Selection

The geometric AIC consists of the residual and the penalty term that is proportional to the degree of freedom of the model. The penalty term is determined by analyzing the decrease of the residual caused by overfitting the model parameters to noisy data [1,4,5]. Adopting the model with the smallest geometric AIC is equivalent to checking how much the residual will increase if the degree of the freedom of the model is reduced and adopting the simpler model *if the resulting increase of the residual is comparable to the decrease of the degree of freedom*, which can be interpreted as a symptom of overfitting.

6 Real Image Experiments

Fig. 5(a) is an image of an outdoor scene. Fig. 5(b) is a zoomed image of the same scene corresponding to the white frame in Fig. 5(a). We manually selected the seven points marked in the images and computed the homography for each of the candidate models described in the preceding section, using the default noise model. The computed geometric AICs of the candidate models are listed in Table 1. As we can see, the similarity model is preferred. Fig. 6(a) shows the resulting superimposed image using the homography computed by the optimal algorithm given in [7]. Fig. 6(b) is the result using the selected similarity.

Fig. 7. Two images of an outdoor scene and the selected points.

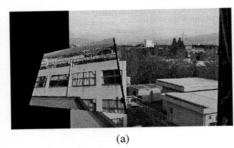 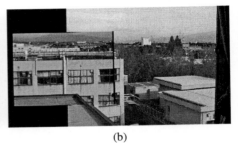

(a) (b)

Fig. 8. (a) Mosaicing by an optimally computed homography. (b) Mosaicing by model selection.

Fig. 7 is a pair of images of an outdoor scene with a small overlap. Using the six points marked in the images, we computed the geometric AICs of the candidate models as shown in Table 1. Again, the similarity model is preferred. Fig. 8(a) is the mosaiced image using the homography computed by the optimal algorithm. Fig. 8(b) is the result using the selected similarity.

Fig. 9 shows a different pair of images. Using the five points marked there, we computed the geometric AICs shown in Table 1, which indicate that the translation model is preferred. Fig. 10(a) is the mosaiced image using the optimal homography; Fig. 10(b) is the result using the selected translation.

Fig. 11 shows the same scene as Fig. 7. This time, we used twenty two points distributed over a large region. The resulting geometric AICs are listed in Table 1. The best model is the 8-parameter homography; the second best model is the 5-parameter homography. The difference between their geometric AICs is very small, indicating that the image transformation can be viewed almost as the 5-parameter homography. Fig. 12(a) is the mosaiced image using the best model; Fig. 12(b) is the result using the second best model.

7 Concluding Remarks

As we can see from Figs. 8(a) and 10(a), the image mapping defined by the optimally computed homography is very unstable and can cause a large unnatural distortion if the

Fig. 9. Two images of an outdoor scene and the selected points.

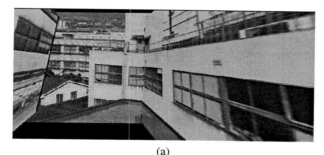

(a)

(b)

Fig. 10. (a) Mosaicing by an optimally computed homography. (b) Mosaicing by model selection.

matching points are small in number and concentrated in a small region in each image. Theoretically, the accuracy cannot be improved any further. We have shown that the accuracy *can* be improved nonetheless if we incorporate our *knowledge* about source of the instability.

The instability stems from the fact that actual transformations of images are usually in a small subgroup of the group of homographies. It follows that undesirable distortions can be removed by selecting an appropriate model by using the geometric AIC. The improvement is dramatic as demonstrated in Figs. 8(b) and 10(b). As Fig. 12 shows, model selection is not necessary if a large number of matching points are distributed over a large region, and the general 8-parameter homography is chosen if model selection

Fig. 11. Matching many points.

 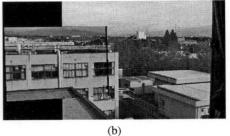

(a)   (b)

Fig. 12. (a) Mosaicing using the best model. (b) Mosaicing using the second best model.

Table 1. The geometric AICs and the selected models.

Model	Fig. 5	Fig. 7	Fig. 9	Fig. 11
8-parameter homography	$9.92E - 06$	$1.25E - 05$	$4.01E - 05$	○ $2.946E - 06$
5-parameter homography	$4.80E - 02$	$3.65E - 03$	$4.69E - 05$	$2.954E - 06$
4-parameter homography	$1.57E - 02$	$4.39E - 02$	$4.45E - 05$	$2.976E - 03$
affine transformation	$8.92E - 06$	$1.08E - 05$	$4.10E - 05$	$3.507E - 06$
similarity	○ $7.32E - 06$	○ $8.54E - 06$	$4.38E - 05$	$4.887E - 06$
rigid motion	$1.57E - 02$	$3.55E - 04$	$4.00E - 05$	$2.976E - 03$
translation	$1.57E - 02$	$3.53E - 04$	○ $3.65E - 05$	$2.990E - 03$

is applied. Thus, an appropriate mapping is always selected whether a sufficient number of matching points are available or not. This selection process does not require any empirical thresholds to adjust. Our technique is very general and can be applied to a wide range of vision applications for increasing accuracy and preventing computational instability (e.g., [8]).

Acknowledgments. This work was in part supported by the Ministry of Education, Science, Sports and Culture, Japan under a Grant in Aid for Scientific Research C(2) (No. 11680377).

References

1. H. Akaike, A new look at the statistical model identification, *IEEE Trans. Automation Control*, **19**-6 (1974), 716–723.
2. W. Förstner, Reliability analysis of parameter estimation in linear models with applications to mensuration problems in computer vision, *Comput. Vision Graphics Image Process.*, **40** (1987), 273–310.
3. K. Kanatani, *Geometric Computation for Machine Vision*, Oxford University Press, Oxford, 1993.
4. K. Kanatani, *Statistical Optimization for Geometric Computation: Theory and Practice*, Elsevier Science, Amsterdam, 1996.
5. K. Kanatani, Geometric information criterion for model selection, *Int. J. Comput. Vision*, **26**-3 (1998), 171–189.
6. K. Kanatani, Statistical optimization and geometric inference in computer vision, *Phil. Trans. Roy. Soc. Lond.*, A-**356** (1998), 1303–1320.
7. K. Kanatani and N. Ohta, Accuracy bounds and optimal computation of homography for image mosaicing applications, *Proc. 7th Int. Conf. Comput. Vision*, September, 1999, Kerkya, Greece, pp. 73–78.
8. C. Matsunaga and K. Kanatani, Calibration of a moving camera using a planar pattern: Optimal computation, reliability evaluation and stabilization by model selection, *Proc. 6th Euro. Conf. Comput. Vision*, June–July 2000, Dublin, Ireland, Vol.2, pp. 595–609.
9. D. D. Morris and T. Kanade, A unified factorization algorithm for points, line segments and planes with uncertainty models, *Proc. Int. Conf. Comput. Vision*, January 1998, Bombay, India, pp. 696–702.
10. H. S. Sawhney, S. Hsu and R. Kumar, Robust video mosaicing through topology inference and local to global alignment, *Proc. 5th Euro. Conf. Comput. Vision*, June 1998, Freiburg, Germany, Vol. 2, pp. 103–119.
11. J. Shi and C. Tomasi, Good features to track, *Proc. Conf. Comput. Vision Patt. Recogn.*, June 1994, Seattle, WA, pp. 593–600.
12. A. Singh, An estimation-theoretic framework for image-flow computation, *Proc. 3rd Int. Conf. Comput. Vision*, December, 1990, Osaka, Japan, pp. 168–177.
13. R. Szeliski and H.-U. Shum, Creating full view panoramic image mosaics and environment maps, *Proc. SIGGRAPH'97*, August 1997, Los Angeles, CA, U.S.A., pp. 251–258.
14. P. H. S. Torr, Model selection for two view geometry: A review, in, D. A. Forsyth, J. L. Mundy, V. D. Gesú, R. Cipolla (Eds.): *Shape, Contour and Grouping in Computer Vision*, LNCS 1681, Springer-Verlag, Berlin, 1999, pp. 277–301.
15. T. Werner, T. Pajdla and V. Hlaváč, Efficient 3-D scene visualization by image extrapolation, *Proc. 5th Euro. Conf. Comput. Vision*, June 1998, Freiburg, Germany, Vol. 2, pp. 382–396.
16. I. Zoghlami, O. Faugeras and R. Deriche, Using geometric corners to build a 2D mosaic from a set of images, *Proc. Conf. Comput. Vision Patt. Recogn.*, June 1997, Puerto Rico, pp. 420–425.

A Analytical Decomposition

We first compute the homography $H = (H_{ij})$ (up to scale) that maps $\{x_\alpha\}$ to $\{x'_\alpha\}$, say, by the optimal algorithm given in [7] or simply by least squares. The non-dimensional focal lengths ϕ' and ϕ and the rotation matrix R that satisfy eq. (11) are computed analytically by the following procedure. First, ϕ' and ϕ are given by

$$\phi' = \sqrt{-\frac{A}{B}}, \qquad \phi = \sqrt{\frac{H_{13}^2 + H_{23}^2 + H_{33}^2 \phi'^2}{K}}, \qquad (30)$$

where

$$A = (H_{11}H_{12} + H_{21}H_{22})H_{31}H_{32} + (H_{12}H_{13} + H_{22}H_{23})H_{32}H_{33}$$
$$+ (H_{13}H_{11} + H_{23}H_{21})H_{33}H_{31}, \qquad (31)$$

$$B = H_{31}^2 H_{32}^2 + H_{32}^2 H_{33}^2 + H_{33}^2 H_{31}^2, \qquad (32)$$

$$K = \frac{H_{11}^2 + H_{21}^2 + H_{12}^2 + H_{22}^2 + (H_{31}^2 + H_{32}^2)\phi'^2}{2}. \qquad (33)$$

Then, compute the following singular value decomposition:

$$\boldsymbol{F}^{-1}\boldsymbol{H}^\top \boldsymbol{F}' = \boldsymbol{V} \begin{pmatrix} \sigma_1 & & \\ & \sigma_2 & \\ & & \sigma_3 \end{pmatrix} \boldsymbol{U}^\top. \qquad (34)$$

Here, $\sigma_1 \geq \sigma_2 \geq \sigma_3 \ (\geq 0)$ are the singular values, and $\boldsymbol{U}$ and $\boldsymbol{V}$ are orthogonal matrices. The rotation matrix $\boldsymbol{R}$ is given by

$$\boldsymbol{R} = \boldsymbol{V} \begin{pmatrix} 1 & & \\ & 1 & \\ & & \det(\boldsymbol{VU}^\top) \end{pmatrix} \boldsymbol{U}^\top. \qquad (35)$$

This procedure produces an exact solution if noise does not exist. In the presence of noise, the solution is optimal in the least squares sense.

B Gradient and Hessian

We put
$$\boldsymbol{e}_\alpha = \boldsymbol{x}'_\alpha \times \boldsymbol{H}\boldsymbol{x}_\alpha. \qquad (36)$$

In computing the gradient ∇J of eq. (7), we ignore terms of $O(\boldsymbol{e}_\alpha)^2$ ($O(\cdots)^n$ denotes terms of order n or higher in $\cdots$). This is justified because J has the form $\sum_{\alpha=1}^{N}(\boldsymbol{e}_\alpha, \boldsymbol{W}_\alpha \boldsymbol{e}_\alpha)/N$ and hence ∇J is $O(\boldsymbol{e}_\alpha)$. In particular, $\boldsymbol{W}_\alpha$ can be regarded as a constant matrix, since the terms involving derivatives of $\boldsymbol{W}$ in ∇J are $O(\boldsymbol{e}_\alpha)^2$. This approximation causes only higher order errors in the solution of $\nabla J = 0$.

Under this approximation, the gradient of J with respect to ϕ, ϕ', and $\boldsymbol{R}$ is given by

$$\frac{\partial J}{\partial \phi} = -\frac{2}{\phi N} \sum_{\alpha=1}^{N} (\boldsymbol{H}\boldsymbol{k}, \boldsymbol{x}'_\alpha \times \boldsymbol{W}_\alpha \boldsymbol{e}_\alpha), \qquad (37)$$

$$\frac{\partial J}{\partial \phi'} = \frac{2}{\phi' N} \sum_{\alpha=1}^{N} (\boldsymbol{k}, \boldsymbol{H}\boldsymbol{x}_\alpha)(\boldsymbol{k}, \boldsymbol{x}'_\alpha \times \boldsymbol{W}_\alpha \boldsymbol{e}_\alpha), \qquad (38)$$

$$\nabla_\Omega J = \frac{2}{N} \sum_{\alpha=1}^{N} (\boldsymbol{F}\boldsymbol{x}_\alpha) \times \boldsymbol{F}^{-1}\boldsymbol{H}^\top (\boldsymbol{x}'_\alpha \times \boldsymbol{W}_\alpha \boldsymbol{e}_\alpha), \qquad (39)$$

where $\boldsymbol{k} = (0, 0, 1)^\top$.

In computing the Hessian of $\nabla^2 J$ of eq. (7), we ignore terms of $O(e_\alpha)$. This is justified because J has the form $\sum_{\alpha=1}^{N}(e_\alpha, W_\alpha e_\alpha)/N$ and hence $\nabla^2 J$ is $O(1)$. In particular, W_α can be regarded as a constant matrix, since the terms involving derivatives of W in $\nabla^2 J$ are $O(e_\alpha)$. This approximation does not affect the accuracy of Newton iterations, since the Hessian $\nabla^2 J$ controls merely the speed of convergence, not the accuracy of the solution. Newton iterations with this approximation, called *Gauss-Newton iterations*, are known to be almost as efficient as Newton iterations.

Under this approximation, the individual elements of the Hessian $\nabla^2 J$ are given as follows:

$$\frac{\partial^2 J}{\partial \phi^2} = \frac{2}{\phi^2 N} \sum_{\alpha=1}^{N} (Hk, (x'_\alpha \times W_\alpha \times x'_\alpha) Hk), \tag{40}$$

$$\frac{\partial^2 J}{\partial \phi'^2} = \frac{2}{\phi'^2 N} \sum_{\alpha=1}^{N} (k, Hx_\alpha)^2 (k, (x'_\alpha \times W_\alpha \times x'_\alpha) k), \tag{41}$$

$$\frac{\partial^2 J}{\partial \phi \partial \phi'} = -\frac{2}{\phi \phi' N} \sum_{\alpha=1}^{N} (k, Hx_\alpha)(Hk, (x'_\alpha \times W_\alpha \times x'_\alpha) k), \tag{42}$$

$$\nabla_\Omega \frac{\partial J}{\partial \phi} = -\frac{2}{\phi N} \sum_{\alpha=1}^{N} (Fx_\alpha) \times F^{-1} H^\top (x'_\alpha \times W_\alpha \times x'_\alpha) Hk, \tag{43}$$

$$\nabla_\Omega \frac{\partial J}{\partial \phi'} = \frac{2}{\phi N} \sum_{\alpha=1}^{N} (k, Hk)(Fx_\alpha) \times F^{-1} H^\top (x'_\alpha \times W_\alpha \times x'_\alpha) k, \tag{44}$$

$$\nabla_\Omega^2 J = \frac{2}{N} \sum_{\alpha=1}^{N} (Fx_\alpha) \times F^{-1} H^\top (x'_\alpha \times W_\alpha \times x'_\alpha) HF^{-1} \times (Fx_\alpha). \tag{45}$$

Here, the product $a \times T \times a$ of a vector $a = (a_i)$ and matrix $T = (T_{ij})$ is a symmetric matrix whose (ij) element is $\sum \varepsilon_{ikl}\varepsilon_{jmn} a_k a_m T_{ln}$, where ε_{ijk} is the Eddington epsilon, taking 1, -1, and 0 when (ijk) is an even permutation of (123), an odd permutation of it, and otherwise, respectively.

C Analytical Similarity Solution

We represent the coordinates (x_α, y_α) and (x'_α, y'_α) and the translation $\vec{\tau} = (\tau_1, \tau_2)^\top$ by the following complex numbers:

$$z_\alpha = \frac{x_\alpha}{f_0} + i\frac{y_\alpha}{f_0}, \quad z'_\alpha = \frac{x'_\alpha}{f_0} + i\frac{y'_\alpha}{f_0}, \quad \tau = \tau_1 + i\tau_2. \tag{46}$$

Let z_C and z'_C be the centroids of the feature points:

$$z_C = \frac{1}{N} \sum_{\alpha=1}^{N} z_\alpha, \quad z'_C = \frac{1}{N} \sum_{\alpha=1}^{N} z'_\alpha. \tag{47}$$

Compute the deviations of the individual feature points from the centroids:

$$\tilde{z}_\alpha = z_\alpha - z_C, \qquad \tilde{z}'_\alpha = z'_\alpha - z'_C. \tag{48}$$

The scale s and the angle θ of rotation are given by

$$s = \frac{1}{N} \sum_{\alpha=1}^{N} \text{abs}[\frac{\tilde{z}'_\alpha}{\tilde{z}_\alpha}], \qquad \theta = \arg[\frac{1}{N} \sum_{\alpha=1}^{N} S[\frac{\tilde{z}'_\alpha}{\tilde{z}_\alpha}]], \tag{49}$$

where we define

$$S[Z] = \frac{Z}{\text{abs}[Z]}. \tag{50}$$

The translation τ is given by

$$\tau = z'_C - se^{i\theta} z_C. \tag{51}$$

Discussion

1. **Fabian Ernst, Philips Research**: You say there is no empirically adjustable threshold involved in your criterion, but you have to make a trade-off between the number of degrees of freedom you have in your homography on the one hand, and residuals on the other hand. Therefore you have implicitly made a trade-off: you have a threshold in the weighting between these two criteria. Could you comment on how sensitive the model selection is to this trade-off?
   Kenichi Kanatani: Yes, we are effectively using some kind of threshold determined by the penalty term for the model complexity, but the penalty term was derived by the theory of Akaike based on statistical principles, not the user. Akaike based his derivation on asymptotic evaluation of the Kullback-Liebler information, but we adopt a different interpretation.
   At any rate, there have been heated arguments among statisticians about how the model complexity should be weighted, and other criteria such as MDL and BIC have also been proposed. In fact, the model selection is a very subtle issue, and we leave it to professionals. If we use other criteria, we may obtain a slightly different result in general. For this mosaicing application, however, we tried other criteria, too, but the result was always the same: the same model was chosen.
2. **Peter Vanroose, Katholieke Universiteit Leuven**: You mention five specific subgroups of the homographies. There are other possible subgroups, did you consider them as well? Would it be worthwhile doing so?
   Kenichi Kanatani: If we would exhaust all possibilities and do model selection, we would end up with something, but this does not make much sense. The success of our method comes from the use of our knowledge that a certain class of transformations is very likely to occur. In this sense, we are implicitly taking the Bayesian approach, since we rely on our prior knowledge about the solution. But we do not explicitly assign any a priori probability to the individual candidate models. I think this is the essence of all techniques using model selection criteria.

3. **Mathias Muehlich, Frankfurt University**: I want to make a comment on your use of the term 'optimal'. You showed that you had to talk about 'optimal' with respect to the model you use, because optimal estimation of your full eight degrees of freedom homography is not optimal for every situation. I would like to add that you should also consider the method you use and the model of errors you use. Because you assume isotropic error, I think that is a rather strong restriction within your model. I would think that if you consider covariances of your input data many strong or severe distortions would not appear. I would not talk about 'optimal' if your renormalization scheme only uses first order approximation. Could you comment on this?

Kenichi Kanatani: Theoretically, the renormalization solution is optimal in the first order. The second order effects are very small, so it is practically optimal. In fact, there exists a theoretical bound beyond which the accuracy cannot be improved, and we have experimentally confirmed that the renormalization solution always falls in the vicinity of that bound.

The next issue is the covariance matrices. Of course, we can adopt anisotropic and inhomogeneous covariance matrices, which can be given by the Hessian of the residual surface of template matching for feature matching. Actually, we did that, but the difference was invisible. We studied the reason carefully. It has turned out that this is because we selected feature points by hand. Humans usually choose very good, salient, features. We do not usually select a point in the sky or on walls of uniform gray levels. If we did, we would have to give such a point a large covariance to compensate for its ambiguity. We also tried automatic feature detectors, but the result was the same. As long as feature detectors or humans eyes are involved, our experience tells us that the assumption of isotropic and homogeneous noise is sufficient and no improvement would result by simply modifying the covariance matrices.

On Computing Metric Upgrades of Projective Reconstructions under the Rectangular Pixel Assumption

Jean Ponce

Dept. of Computer Science and Beckman Institute
University of Illinois at Urbana-Champaign, USA

Abstract. This paper shows how to upgrade the projective reconstruction of a scene to a metric one in the case where the only assumption made about the cameras observing that scene is that they have rectangular pixels (zero-skew cameras). The proposed approach is based on a simple characterization of zero-skew projection matrices in terms of line geometry, and it handles zero-skew cameras with arbitrary or known aspect ratios in a unified framework. The metric upgrade computation is decomposed into a sequence of linear operations, including linear least-squares parameter estimation and eigenvalue-based symmetric matrix factorization, followed by an optional non-linear least-squares refinement step. A few classes of critical motions for which a unique solution cannot be found are spelled out. A MATLAB implementation has been constructed and preliminary experiments with real data are presented.

1 Introduction

The past ten years have witnessed very impressive progress in motion analysis. Keys to this progress have been the emergence of reliable interest-point detectors (e.g., [11]) and feature trackers (e.g., [31]); a shift from methods relying on a minimum number of images (e.g., [33]) to techniques using a large number of pictures (e.g., [24,30,31]), facilitated by the decrease in price of image acquisition and storage hardware; and a vastly improved understanding of the geometric, statistical and numerical issues involved (e.g., [5,6,14,15,19,24,30,31]). For example, Tomasi and Kanade [30] and their colleagues [2,22] have shown that the motion of a calibrated orthographic, weak perspective or paraperspective image can be estimated by first using singular value decomposition to compute an affine reconstruction of the observed scene, then upgrading this reconstruction to a full metric one using the Euclidean constraints available from the calibration parameters [19,25]. We consider in this paper the more complicated case of perspective projection, where n fixed points P_j ($j = 1, \ldots, n$) are observed by m perspective cameras. Given some fixed world coordinate system, we can write

$$p_{ij} = \mathcal{M}_i P_j \quad \text{for} \quad i = 1, \ldots, m \quad \text{and} \quad j = 1, \ldots, n, \tag{1}$$

where p_{ij} denotes the (homogeneous) coordinate vector of the projection of the point j in the image i expressed in the corresponding camera's coordinate system, $\mathcal{M}_i$ is the

3×4 projection matrix associated with this camera in the world coordinate system, and $\boldsymbol{P}_j$ is the homogeneous coordinate vector of the point P_j in that coordinate system.

We address the problem of reconstructing both the matrices $\mathcal{M}_i$ ($i = 1, \ldots, m$) and the vectors $\boldsymbol{P}_j$ ($j = 1, \ldots, n$) from the image correspondences $\boldsymbol{p}_{ij}$. Faugeras [5] and Hartley et al. [15] have shown that when no assumption is made about the internal parameters of the cameras, such a reconstruction can only be done up to an arbitrary projective transformation, i.e., if $\mathcal{M}_i$ and $\boldsymbol{P}_j$ are solutions of (1), so are $\mathcal{M}_i \mathcal{Q}$ and $\mathcal{Q}^{-1} \boldsymbol{P}_j$ for any nonsingular 4×4 matrix $\mathcal{Q}$. Several effective techniques for computing a projective scene representation from multiple images have been proposed (e.g., [5,12,15,21,27,28,35]). As in the affine case, the projective reconstruction can be upgraded to a full metric model [7] by exploiting a priori knowledge of camera calibration parameters (e.g., [8,9,13,20,23,24]) or scene geometry (e.g., [1]).

Now, although the internal parameters of a camera may certainly be unknown (e.g., when stock footage is used) or change from one image to the next (e.g., when several different cameras are used to film a video clip, or when a camera zooms, which will change both its focal length and the position of its principal point), there is one parameter that will, in practice, never change: it is the skew of the camera, i.e., the difference between $\pi/2$ and the angle actually separating the rows and columns of an image. Except possibly for minute manufacturing errors, the skew will always be zero. Likewise, the aspect ratio of a camera will never change, and it may be known a priori. Zero-skew perspective projection matrices have been characterized by Faugeras [6, Theorems 3.1 and 3.2] and Heyden [16] as follows.

Lemma 1. *A necessary and sufficient condition for a rank-3 3×4 matrix*

$$\mathcal{M} = \begin{pmatrix} \boldsymbol{m}_1^T & m_{14} \\ \boldsymbol{m}_2^T & m_{24} \\ \boldsymbol{m}_3^T & m_{34} \end{pmatrix}$$

to be a zero-skew perspective projection matrix is that

$$(\boldsymbol{m}_1 \times \boldsymbol{m}_3) \cdot (\boldsymbol{m}_2 \times \boldsymbol{m}_3) = 0, \tag{2}$$

and a necessary and sufficient condition for a zero-skew perspective projection matrix $\mathcal{M}$ to have unit aspect ratio is that

$$|\boldsymbol{m}_1 \times \boldsymbol{m}_3|^2 = |\boldsymbol{m}_2 \times \boldsymbol{m}_3|^2. \tag{3}$$

Let us follow Faugeras [6] and give a geometric interpretation of this lemma: the rows of the matrix $\mathcal{M}$ are associated with the planes $\Pi_i : \boldsymbol{m}_i \cdot \boldsymbol{x} + m_{i4} = 0$ ($i = 1, 2, 3$), called *projection planes* in [4]. The image coordinate axis $u = 0$ of the image is parallel to the line λ where Π_1 intersects the focal plane (i.e., the plane parallel to the retina that passes through the optical center) Π_3, and its direction is the cross product $\boldsymbol{m}_1 \times \boldsymbol{m}_2$ of the two plane normals. Likewise, the coordinate axis $v = 0$ is parallel to the line $\mu = \Pi_2 \cap \Pi_3$ and its direction is $\boldsymbol{m}_2 \times \boldsymbol{m}_3$. Equation (2) simply expresses the fact that these two lines are perpendicular. The additional condition in (3) expresses the fact that the scales of the two image coordinate axes are the same.

Lemma 1 shows that arbitrary 3×4 matrices are not zero-skew perspective projection matrices. It can therefore be hoped that better-than-projective reconstructions of the world can be achieved for zero-skew cameras (and a fortiori for cameras with zero skew and unit aspect ratio). We will say that a projective transformation $\mathcal{Q}$ preserves zero skew when, for *any* zero-skew perspective projection matrix $\mathcal{M}$, the matrix $\mathcal{MQ}$ is also a zero-skew perspective projection matrix. Heyden and Åström [17] and Pollefeys *et al.* [24] have independently shown the following important result.

Lemma 2. *The class of transformations that preserve zero skew is the group of similarity transformations.*

Similarity transformations obviously preserve the aspect ratio of a camera so the above result also holds for zero-skew cameras with unit aspect ratio.

The proof of this lemma is constructive: for example, Pollefeys *et al.* [24] exhibit a set of eight camera positions and orientations that constrain the transformation to be a similarity. In this setting, Heyden and Åström [18] have also given a bundle-adjustment method for estimating the calibration parameters as well as the metric structure and motion parameters. Their method is not linear and it relies on the algorithm proposed by Pollefeys *et al.* [24] to find an initial guess assuming known principal point and aspect ratio. We use line geometry to derive in the rest of this paper a quasi-linear alternative to that technique that does not require any initial guess and handles both arbitrary zero-skew matrices and zero-skew matrices with unit aspect ratio in a unified framework. In addition, we spell out a few classes of critical motions for which a unique solution cannot be found, and present some preliminary experiments with real data.

2 A Characterization of Metric Upgrades for Zero-Skew Cameras

Suppose that some projective reconstruction technique (e.g., [5,15]) has been used to estimate the projection matrices $\mathcal{M}_i$ ($i = 1, \ldots, m$) and the point positions $\boldsymbol{P}_j$ ($j = 1, \ldots, n$) from m images of these points. We know that any other reconstruction *and in particular a metric one* will be separated from this one by a projective transformation. This section provides an algebraic and geometric characterization of the 4×4 matrices $\mathcal{Q}$ such that, if $\hat{\mathcal{M}} = \mathcal{MQ}$, the rows of $\hat{\mathcal{M}}$ satisfy the condition of Lemma 2. These transformations are called zero-skew metric upgrades in the sequel. To characterize these transformations in a simple manner, it is useful to recall some elementary notions of line geometry (see [3,4] for related applications to motion analysis). Let us first introduce the operator "$\wedge$" that associates with two 4-vectors $\boldsymbol{a}$ and $\boldsymbol{b}$ their *exterior product* defined as the 6-vector

$$\boldsymbol{a} \wedge \boldsymbol{b} \stackrel{\text{def}}{=} \begin{pmatrix} a_1 b_2 - a_2 b_1 \\ a_1 b_3 - a_3 b_1 \\ a_1 b_4 - a_4 b_1 \\ a_2 b_3 - a_3 b_2 \\ a_2 b_4 - a_4 b_2 \\ a_3 b_4 - a_4 b_3 \end{pmatrix}.$$

Note the similarity with the cross product operator that also associates with two vectors (3-vectors of course, instead of 4-vectors) $\boldsymbol{a}$ and $\boldsymbol{b}$ the vector formed by all the 2×2 minors of the matrix $(\boldsymbol{a}, \boldsymbol{b})$.

Geometrically, the exterior product associates with the homogeneous coordinate vectors of two points in $\mathbf{P}^3$ the so-called *Plücker coordinates* of the line joining them. In a dual sense, it also associates with two planes in $\mathbf{P}^3$ the line where these planes intersect. Plücker coordinates are homogeneous and lines form a subspace of dimension 4 of the projective space $\mathbf{P}^5$: indeed, it follows immediately from the definition of the exterior product that the Plücker coordinate vector $l = (l_1, l_2, l_3, l_4, l_5, l_6)^T$ of a line obeys the quadratic constraint

$$l_1 l_6 - l_2 l_5 + l_3 l_4 = 0. \tag{4}$$

It is also possible to define an inner product on the set of all lines by the formula

$$(l|l') \stackrel{\text{def}}{=} l_1 l'_6 + l_6 l'_1 - l_2 l'_5 - l_5 l'_2 + l_3 l'_4 + l_4 l'_3.$$

Clearly, a 6-vector l represents a line if and only if $(l|l) = 0$, and it can also be shown that a necessary and sufficient condition for two lines to be coplanar is that $(l|l') = 0$. It will also prove convenient in the sequel to define the vector $\bar{l} = (l_6, -l_5, l_4, l_3, -l_2, l_1)^T$, so that $(l|l') = l^T \bar{l}' = \bar{l}^T l'$.

We are now in a position to characterize zero-skew metric upgrades. We write the matrices $\hat{\mathcal{M}}$, $\mathcal{M}$ and $\mathcal{Q}$ as

$$\hat{\mathcal{M}} = \begin{pmatrix} \hat{m}_1^T & \hat{m}_{14} \\ \hat{m}_2^T & \hat{m}_{24} \\ \hat{m}_3^T & \hat{m}_{34} \end{pmatrix}, \quad \mathcal{M} = \begin{pmatrix} m_1^T \\ m_2^T \\ m_3^T \end{pmatrix} \quad \text{and} \quad \mathcal{Q} = \begin{pmatrix} q_1 & q_2 & q_3 & q_4 \end{pmatrix}.$$

Note that the vectors m_i and q_i are elements of $\mathbf{R}^4$ but the vectors $\hat{m}_i$ are elements of $\mathbf{R}^3$. With this notation, we have the following result.

Lemma 3. *Given a projection matrix $\mathcal{M}$ and a projective transformation $\mathcal{Q}$, a necessary and sufficient condition for the matrix $\hat{\mathcal{M}} = \mathcal{M}\mathcal{Q}$ to satisfy the zero-skew constraint*

$$(\hat{m}_1 \times \hat{m}_3) \cdot (\hat{m}_2 \times \hat{m}_3) = 0$$

is that

$$\lambda^T \mathcal{R}^T \mathcal{R} \mu = 0, \tag{5}$$

where

$$\mathcal{R} \stackrel{\text{def}}{=} \begin{pmatrix} (q_2 \wedge q_3)^T \\ (q_3 \wedge q_1)^T \\ (q_1 \wedge q_2)^T \end{pmatrix}, \quad \lambda \stackrel{\text{def}}{=} m_1 \wedge m_3 \quad \text{and} \quad \mu \stackrel{\text{def}}{=} m_2 \wedge m_3.$$

In addition, a necessary and sufficient condition for the zero-skew perspective projection matrix $\hat{\mathcal{M}}$ to have unit aspect ratio is that

$$\lambda^T \mathcal{R}^T \mathcal{R} \lambda = \mu^t \mathcal{R}^T \mathcal{R} \mu. \tag{6}$$

The proof of this lemma relies on elementary properties of the exterior product to show that $\hat{m}_1 \times \hat{m}_3 = \mathcal{R}\lambda$ and $\hat{m}_2 \times \hat{m}_3 = \mathcal{R}\mu$, from which the result immediately follows. Its geometric interpretation is simple as well: obviously, the vector λ is the vector of Plücker coordinates of the line λ formed by the intersection of the planes Π_1 and Π_3 associated with the projection matrix $\mathcal{M}$. Likewise, μ is the vector of Plücker coordinates of the line $\mu = \Pi_2 \cap \Pi_3$. When the transformation $\mathcal{Q}$ is applied to the matrix $\mathcal{M}$, these two lines map onto lines $\hat{\lambda}$ and $\hat{\mu}$ parallel to the coordinate axes $\hat{u} = 0$ and $\hat{v} = 0$ of the zero-skew image. As shown in Appendix A, the matrix $\mathcal{R}$ maps lines onto the direction of their image under $\mathcal{Q}$, thus (5) simply expresses the fact that the lines $\hat{u} = 0$ and $\hat{v} = 0$ are perpendicular. As before, the additional condition (6) expresses that the scales of the two image coordinate axes are the same.

3 Computing the Upgrade

We show in this section that a matrix $\mathcal{Q}$ satisfying (5) can be estimated from at least 19 images using linear methods: we first use linear least squares to estimate the matrix $\mathcal{S} \stackrel{\text{def}}{=} \mathcal{R}^T \mathcal{R}$, then take advantage of elementary properties of symmetric (but possibly indefinite) matrices to factor $\mathcal{S}$ and compute $\mathcal{R}$. Once $\mathcal{R}$ is known, it is a simple matter to determine the matrix $\mathcal{Q}$ using once again linear least squares. The proposed approach linearizes the estimation process since (5) is an equation of degree 4 in the coefficients of $\mathcal{Q}$. The following lemma clarifies the corresponding properties of the matrices $\mathcal{R}$ and $\mathcal{S}$.

Lemma 4. *The matrices $\mathcal{R}$ and $\mathcal{S}$ have the following properties:*

1. *The columns $\boldsymbol{R}_1$, $\boldsymbol{R}_2$ and $\boldsymbol{R}_3$ of the matrix $\mathcal{R}^T$ satisfy the 6 quadratic constraints*

$$\begin{cases} (\boldsymbol{R}_1 | \boldsymbol{R}_1) = 0, \\ (\boldsymbol{R}_2 | \boldsymbol{R}_2) = 0, \\ (\boldsymbol{R}_3 | \boldsymbol{R}_3) = 0, \end{cases} \text{ and } \begin{cases} (\boldsymbol{R}_1 | \boldsymbol{R}_2) = 0, \\ (\boldsymbol{R}_2 | \boldsymbol{R}_3) = 0, \\ (\boldsymbol{R}_3 | \boldsymbol{R}_1) = 0. \end{cases}$$

2. *The coefficients S_{ij} of the matrix $\mathcal{S}$ satisfy the linear constraint*

$$S_{16} - S_{25} + S_{34} = 0.$$

3. *The columns $\boldsymbol{S}_i$ $(i = 1, \ldots, 6)$ of $\mathcal{S}$ satisfy the 12 quadratic constraints*

$$\begin{cases} (\boldsymbol{S}_1 | \boldsymbol{S}_1) = 0, \\ (\boldsymbol{S}_1 | \boldsymbol{S}_2) = 0, \\ (\boldsymbol{S}_1 | \boldsymbol{S}_3) = 0, \\ (\boldsymbol{S}_2 | \boldsymbol{S}_2) = 0, \\ (\boldsymbol{S}_2 | \boldsymbol{S}_3) = 0, \\ (\boldsymbol{S}_3 | \boldsymbol{S}_3) = 0, \end{cases} \text{ and } \begin{cases} (\boldsymbol{S}_4 | \boldsymbol{S}_1) = 0, \\ (\boldsymbol{S}_4 | \boldsymbol{S}_2) = 0, \\ (\boldsymbol{S}_4 | \boldsymbol{S}_3) = 0, \\ (\boldsymbol{S}_5 | \boldsymbol{S}_1) = 0, \\ (\boldsymbol{S}_5 | \boldsymbol{S}_2) = 0, \\ (\boldsymbol{S}_6 | \boldsymbol{S}_1) = 0. \end{cases}$$

The proof of this lemma is simple and it can be found in Appendix B. It relies on showing that the columns of these two matrices are the Plücker coordinates of a certain number of lines. Note that the quadratic constraints satisfied by the entries of the matrix $\mathcal{S}$ capture the linear dependency of its columns and the fact that it has (at most) rank 3.

3.1 Computing $\mathcal{S}$: Linear Least Squares

We are now ready to present our method for estimating the matrices $\mathcal{S}$, $\mathcal{R}$ and $\mathcal{Q}$ associated with zero-skew cameras. Let us first note that (5) is a linear constraint on the coefficients of $\mathcal{S}$, that can be rewritten as

$$\sum_{i=1}^{6} \lambda_i \mu_i S_{ii} + \sum_{1 \leq i < j \leq 6} (\lambda_i \mu_j + \lambda_j \mu_i) S_{ij} = 0 \qquad (7)$$

where the coefficients λ_i and μ_i denote the coordinates of the vectors $\boldsymbol{\lambda}$ and $\boldsymbol{\mu}$ and the 20 coefficients S_{ij} denote the entries of $\mathcal{S}$.

According to Property 2 in Lemma 4, we have $S_{16} - S_{25} + S_{34} = 0$. In addition, since the lines associated with the vectors $\boldsymbol{\lambda}$ and $\boldsymbol{\mu}$ both lie in the focal plane, we have $(\boldsymbol{\lambda}|\boldsymbol{\mu}) = 0$. This allows us to eliminate the unknown S_{16} and rewrite (7) as

$$\sum_{i=1}^{6} \lambda_i \mu_i S_{ii} + \sum_{\substack{1 \leq i < j \leq 6 \\ i+j \neq 7}} (\lambda_i \mu_j + \lambda_j \mu_i) S_{ij} + a_{25} S_{25} + a_{34} S_{34} = 0, \qquad (8)$$

where

$$\begin{cases} a_{25} = 2(\lambda_2 \mu_5 + \lambda_5 \mu_2) - (\lambda_3 \mu_4 + \lambda_4 \mu_3), \\ a_{34} = 2(\lambda_3 \mu_4 + \lambda_4 \mu_3) - (\lambda_2 \mu_5 + \lambda_5 \mu_2), \end{cases}$$

and the missing elements in the second sum in (8) correspond to the terms S_{16}, S_{25} and S_{34}.

With only 20 out of the 21 original unknown coefficients left, writing (8) for $m \geq 19$ images yields an overdetermined homogeneous system of linear equations of the form $\mathcal{A}s = 0$, where $\mathcal{A}$ is an $m \times 20$ data matrix and s is the vector formed by the 20 independent coefficients of $\mathcal{S}$. The least-squares solution of this system is computed (up to an irrelevant scale factor) as the rightmost column of the 20×20 matrix $\mathcal{V}$ in the singular value decomposition $\mathcal{UDV}^T$ of $\mathcal{A}$. The S_{16} entry is then computed as $S_{25} - S_{34}$. Note that this linear process ignores the 12 quadratic equations satisfied by the entries of the matrix $\mathcal{S}$ according to Lemma 4. This suggests a two-pass estimation process, where the coefficients of $\mathcal{S}$ are first estimated using linear least squares and then refined using constrained optimization.

The method is readily adapted to the case of zero-skew matrices with unit (or, equivalently, known) aspect ratio by adding to the linear constraint (8) associated with (5) the linear constraint

$$\sum_{i=1}^{6} (\lambda_i^2 - \mu_i^2) S_{ii} + \sum_{\substack{1 \leq i < j \leq 6 \\ i+j \neq 7}} 2(\lambda_i \lambda_j - \mu_i \mu_j) S_{ij} + 2b_{25} S_{25} + 2b_{34} S_{34} = 0 \qquad (9)$$

associated with (6), where

$$\begin{cases} b_{25} = 2(\lambda_2 \lambda_5 - \mu_2 \mu_5) - (\lambda_3 \lambda_4 - \mu_3 \mu_4), \\ b_{34} = 2(\lambda_3 \lambda_4 - \mu_3 \mu_4) - (\lambda_2 \lambda_5 - \mu_2 \mu_5). \end{cases}$$

3.2 Computing $\mathcal{R}$: Factorization of Symmetric Matrices

In both cases, once the symmetric matrix $\mathcal{S}$ is known, it can be used to estimate the matrix $\mathcal{R}$: for example, if $\mathcal{S}$ is positive (semidefinite by construction in the noiseless case, but possibly definite in the presence of noise), its singular value decomposition has the form $\mathcal{S} = \mathcal{U}\mathcal{D}\mathcal{U}^T$ and $\mathcal{R}^T$ can be taken equal to $\mathcal{U}_3\sqrt{\mathcal{D}_3}$, where $\mathcal{U}_3$ is the matrix formed by the three columns of $\mathcal{U}$ associated with the largest singular values of $\mathcal{S}$. This construction relies on the well-known fact that the closest rank-3 approximation to a given matrix in the sense of the Frobenius form is obtained by zeroing its three smallest singular values, and it has been used in various contexts in computer vision (e.g., [14,32,34]).

Unfortunatey, in the presence of noise, $\mathcal{S}$ is not guaranteed (and in fact is unlikely) to be positive, and the above method does not apply (see, for example, [25,34] for a discussion of this problem). To tackle this difficulty, we will use an elementary property of symmetric matrices: let us consider an arbitrary $n \times n$ symmetric matrix $\mathcal{S}$ with real coefficients, and diagonalize this matrix in an orthonormal basis as $\mathcal{S} = \mathcal{U}\mathcal{D}\mathcal{U}^T$, where $\mathcal{D}$ is the diagonal matrix formed by the (possibly negative) eigenvalues of $\mathcal{S}$ and $\mathcal{U}$ is the orthogonal matrix formed by its eigenvectors. We seek the $n \times n$ symmetric positive semidefinite matrix $\hat{\mathcal{S}}$ that best approximates $\mathcal{S}$ in the sense of the Frobenius form, i.e., minimizes

$$E^2 = |\mathcal{S} - \hat{\mathcal{S}}|_2^2 = \sum_{i,j=1}^{n}(S_{ij} - \hat{S}_{ij})^2.$$

Property 1. The symmetric definite semipositive matrix $\hat{\mathcal{S}}$ minimizing E^2 is $\mathcal{U}\mathcal{D}_0\mathcal{U}^T$, where $\mathcal{D}_0$ is the diagonal matrix obtained by setting all negative entries of $\mathcal{D}$ to zero.

The proof of this property is simple and it is given in Appendix C.[1]

In our setting, we first compute the eigenvectors and eigenvalues of $\mathcal{S}$, then zero the negative eigenvalues. At this point it is still possible because of noise that more than three of the eigenvalues be positive. To enforce the rank-3 constraint we use the property of singular value decomposition mentioned before and zero all remaining eigenvalues but the three largest ones. This step is justified by the fact that the singular values of a symmetric matrix are the absolute values of its eigenvalues. Finally, we set $\mathcal{R}^T = \mathcal{U}_3\sqrt{\mathcal{D}_3}$, where $\mathcal{U}_3$ is the matrix formed by the columns of $\mathcal{U}$ associated with the remaining eigenvalues of $\mathcal{S}$.

Note that this process only determines $\mathcal{R}$ up to an arbitrary 3×3 orthogonal matrix $\mathcal{A}$ since, if $\mathcal{S} = \mathcal{R}^T\mathcal{R}$, then we also have $\mathcal{S} = \mathcal{R}'^T\mathcal{R}'$, where $\mathcal{R}' = \mathcal{A}\mathcal{R}$. Conversely, although $\mathcal{R}$ can only be estimated up to an arbitrary orthogonal transformation $\mathcal{A}$, the coefficients of the matrix $\mathcal{S}$ are by construction invariant under $\mathcal{A}$. It should also be noted that this factorization approach ignores the 6 quadratic constraints satisfied by the entries of the matrix $\mathcal{R}$ according to Lemma 4. Again, this suggests a two-pass process using the result of factorization as a seed for a second constrained optimization stage.

---

[1] This is for completeness only since we have not been able to find the appropriate reference yet. It should be noted that optimization algorithms routinely rely on positive definite approximations of indefinite symmetric matrices to improve the numerical stability of their output (e.g., [10,26]). The problem is a bit different here since we seek a positive *semidefinite* approximation.

3.3 Computing $\mathcal{Q}$: Linear Least Squares

Once $\mathcal{R}$ is known, we can estimate the vectors q_1, q_2 and q_3 using linear least squares thanks to the following classical property of Plücker coordinates, given here without proof.

Property 2. Given a line l with Plücker coordinate vector $l = (l_1, l_2, l_3, l_4, l_5, l_6)^T$ and a point (resp. plane) P with homogeneous coordinate vector P, a necessary and sufficient condition for P to lie on l (resp. for l to lie in P) is that

$$\mathcal{L}P = 0, \quad \text{where} \quad \mathcal{L} \stackrel{\text{def}}{=} \begin{pmatrix} 0 & l_6 & -l_5 & l_4 \\ -l_6 & 0 & l_3 & -l_2 \\ l_5 & -l_3 & 0 & l_1 \\ -l_4 & l_2 & -l_1 & 0 \end{pmatrix},$$

and the plane Π spanned by the line l and the point P (resp. the point Π where the line l and the plane P intersect) has homogeneous coordinates $\Pi = \mathcal{L}P$.

This result allows us to write constraints such as $\mathcal{L}_{12}q_1 = 0$ and $\mathcal{L}_{12}q_3 = \mathcal{L}_{23}q_1$, where $\mathcal{L}_{ij}$ denotes the $\mathcal{L}$ matrix associated with the estimated value of $q_i \wedge q_j$ for $i, j = 1, 2, 3$.[2] Collecting the $6 \times 3 + 3 \times 4 = 30$ different equations of this type obtained by permuting the appropriate subscripts yields a systems of linear equations in the coordinates of the vectors q_i that can be solved once again using linear least squares (at most 11 of the 30 equations are independent in the noise-free case).

Once the vectors q_i are known, we can complete the construction of $\mathcal{Q}$ by imposing, for example, that the optical center of the first camera be used as origin of the world coordinate system. This translates into the fourth column of $\hat{\mathcal{M}}_1$ being zero, and allows us to compute q_4 (up to scale) as the solution of $\mathcal{M}_1 q_4 = 0$. This unknown scale factor reflects the fact that we have a metric but not Euclidean reconstruction, i.e., absolute scale cannot be recovered.

3.4 Refining $\mathcal{Q}$: Non-Linear Least Squares

Let us conclude by noting that, given m projection matrices $\mathcal{M}_i$, the estimates of the vectors q_i ($i = 1, 2, 3$) obtained from the linear least-squares process can be refined using non-linear least-squares to minimize the average squared skew of the projection matrices, i.e.,

$$\frac{1}{m} \sum_{i=1}^{m} \left[\arcsin \frac{(\mathcal{R}\lambda_i) \cdot (\mathcal{R}\mu_i)}{|\mathcal{R}\lambda_i| |\mathcal{R}\mu_i|} \right]^2, \tag{10}$$

with respect to the vectors q_i ($i = 1, 2, 3$). The vector q_4 can then be computed as before. We have implemented this method and present a comparison with linear least squares in Section 5.

---

[2] This is true despite the fact that the homogeneous coordinate vector Π in Property 2 is only defined up to scale: it is indeed easy to show that we can write $\mathcal{L}_{12}q_3 = \mathcal{L}_{23}q_1$ instead of $\mathcal{L}_{12}q_3 = \rho \mathcal{L}_{23}q_1$ because of the particular method used to construct the vectors L_{ij}.

4 Degenerate Motions

It is of course important to understand the conditions under which the proposed method will fail. Let us consider first the case of arbitrary zero-skew cameras. We assume that our data consist of $m \geq 19$ matrices $\mathcal{M}_i$ $(1 = 1, \ldots, m)$ and denote by λ_i and μ_i the associated vectors of Plücker coordinates. The matrix $\mathcal{S}$ we seek is a solution of the linear system of equations

$$\lambda_i^T \mathcal{X} \mu_i = 0 \quad \text{for} \quad i = 1, \ldots, m. \tag{11}$$

The linear least-squares estimation of $\mathcal{S}$ will fail when the associated $m \times 20$ data matrix has rank less than 19, or, more directly, when (11) does not admit a unique solution.

The equation $\lambda^T \mathcal{X} \mu = 0$ defines a quadric surface in the space spanned by the vectors λ and μ and a quartic surface in the space of all projection matrices. When $\mathcal{X}$ is equal to $\mathcal{S}$, this quartic surface is precisely the 10-dimensional set of all zero-skew projection matrices. For a motion sequence to be degenerate, the projection matrices must lie on a second quartic surface as well, which will never occur for general enough motions. When the camera motion is sufficiently restricted, however, (11) may admit several solutions. Identifying all possible cases is difficult, but we can spell out a few simple ones. Suppose for instance that there exists some fixed line ξ such that λ and ξ remain coplanar during the whole motion sequence. In this case we obviously have $(\lambda | \xi) = \lambda^T \bar{\xi} = 0$, thus $\lambda^T \mathcal{X} \mu = 0$ with $\mathcal{X} = \bar{\xi}\bar{\xi}^T$, and the method will fail. The same is of course true when there exists a fixed line ξ such that ξ and μ are coplanar for every image in the sequence. The following lemma identifies a few classes of such degenerate motion sequences.

Lemma 5. *The following classes of motions of an arbitrary zero-skew camera do not determine a unique metric reconstruction (independently of the estimation method actually used):*

1. *Pure translations: the optical center of the camera may change in an arbitrary manner but the camera's orientation is held constant.*
2. *Planar motions: the optical center is held in the plane $y = 0$ and the camera is allowed to rotate about the y axis.*
3. *Straight-line motions: the optical center of the camera moves along a straight line but the orientation of the camera is allowed to change arbitrarily.*

These are well-known degenerate motions for several self-calibration methods (e.g., [23,29,31]). Note that straight-line motions include pure rotations. The lemma is proven by choosing an appropriate line ξ for each motion class: for pure translations, the image coordinate axes translate parallel to themselves, and we can pick ξ to be some fixed line parallel to λ or to μ. For planar motions, the line μ remains in the plane $y = 0$ and we can pick any fixed line in this plane for ξ. In the case of a straight-line optical center motion, we can pick ξ to be the trajectory of the optical center, since it will always intersect both λ and μ. Note that the motions identified by Lemma 5 will remain degenerate even if we impose that the entries of the matrix $\mathcal{S}$ satisfy the 12 quadratic constraints of Lemma 4:

indeed, the columns of $\bar{\xi}\bar{\xi}^T$ are scaled versions of the same Plücker coordinate vector and they satisfy these constraints.

The case of zero-skew cameras with unit aspect ratio is a bit different since in this case ξ must intersect (or be parallel to) *both* λ and μ. In particular translations and planar motions are not obviously degenerate motions in this case (they still may be since the existence of the line ξ intersecting λ and μ is only a *sufficient* condition for degeneracy), but straight-line motions remain degenerate. Additional work is needed to give necessary and sufficient conditions for degeneracy.

5 Implementation and Results

A preliminary MATLAB implementation of the proposed approach has been constructed, and tested with real data kindly provided by Marc Pollefeys. The linear least-squares estimation of S and Q has been implemented by the MATLAB `svd` routine for singular value decomposition. The factorization of S has been implemented using the `eig` function for eigenvalue and eigenvector computation, and the `lsqnonlin` routine has been used to perform the non-linear least-squares refinement of Q. The constrained optimization processes for estimating S and $\mathcal{R}$ mentioned in Section 3 have not been implemented. Our data consist of projective reconstructions of 182 projection matrices and 3506 points from a sequence of images of a desk scene featuring a volleyball and a cylindrical box. We have assumed in our experiments that all cameras have zero skew but arbitrary aspect ratio.

Figure 1 shows our results, including plots of the original projective reconstruction (Figure 1(a)), the metric reconstruction obtained using the self-calibration method proposed by Pollefeys *et al.* [23,24] (Figure 1(b)), and the metric reconstructions using our method and both linear least squares (Figure 1(c)) and non-linear optimization (Figure 1(d)). These results are a bit difficult to evaluate objectively since (1) ground truth is not available, (2) the data points in the metric reconstruction of Pollefeys *et al.* are sampled quite differently from those used in the projective reconstruction and our metric upgrades, and (3) the results are not shown from the same viewpoints (due to the facts that the reconstruction is only done up to an arbitrary rigid transformation plus scaling and that we have not yet implemented an automatic registration program). Still, the two parallel planes and the spherical shape of the ball seem to be rather well preserved in our reconstructions. The linear estimation of Q takes 0.5s on a Pentium II 450MHz machine, and yields an average skew of $5.68°$ over the 182 input matrices. Starting from the linear estimate, the non-linear least-squares function `lsqnonlin` converges in 9s after 16 iterations and yields an average skew of $0.46°$. More experiments are of course necessary to validate our approach.

Acknowledgments. I would like to thank Marc Pollefeys for providing the data used in our experiments, and Mike Heath, Martial Hebert, Seth Hutchinson, David Kriegman, Pierre Moulin, Bob Skeel and Eric de Sturler for useful discussions and comments.

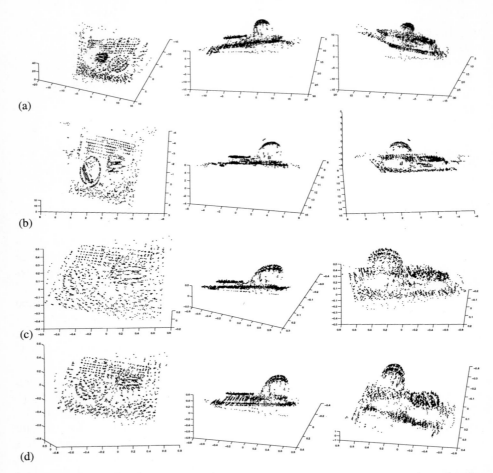

Fig. 1. Experimental results: (a) projective reconstruction; (b) metric reconstruction using the method described in [23,24]; (c) metric reconstruction obtained by the method presented in this paper using linear least squares; (d) metric reconstruction using non-linear least squares.

Appendix

Appendix A: Proof of Lemma 3

Let us consider a line l defined by the intersection of two arbitrary planes with coordinate vectors m and n. The Plücker coordinate vector l of this line is equal to $m \wedge n$, and its image under the transformation $\mathcal{Q}$ is

$$\hat{l} = (\mathcal{Q}^T m) \wedge (\mathcal{Q}^T n) = \begin{pmatrix} (q_1 \cdot m)(q_2 \cdot n) - (q_2 \cdot m)(q_1 \cdot n) \\ (q_1 \cdot m)(q_3 \cdot n) - (q_3 \cdot m)(q_1 \cdot n) \\ (q_1 \cdot m)(q_4 \cdot n) - (q_4 \cdot m)(q_1 \cdot n) \\ (q_2 \cdot m)(q_3 \cdot n) - (q_3 \cdot m)(q_2 \cdot n) \\ (q_2 \cdot m)(q_4 \cdot n) - (q_4 \cdot m)(q_2 \cdot n) \\ (q_3 \cdot m)(q_4 \cdot n) - (q_4 \cdot m)(q_2 \cdot n) \end{pmatrix}.$$

Now, note that the direction of any line l is the cross product of the normals of the two planes defining it. In other words, if $l = (l_1, l_2, l_3, l_4, l_5, l_6)^T$ is the line's vector of Plücker coordinates, then its direction is $v = (l_4, -l_2, l_1)^T$. Applying this result to the line $\hat{l}$ yields

$$\hat{v} = \begin{pmatrix} (q_2 \cdot m)(q_3 \cdot n) - (q_3 \cdot m)(q_2 \cdot n) \\ (q_3 \cdot m)(q_1 \cdot n) - (q_1 \cdot m)(q_3 \cdot n) \\ (q_1 \cdot m)(q_2 \cdot n) - (q_2 \cdot m)(q_1 \cdot n) \end{pmatrix}. \quad (12)$$

It is easy to check analytically that the following identity holds for any 4-vectors a, b, c and d:

$$(a \wedge b) \cdot (c \wedge d) = (a \cdot c)(b \cdot d) - (a \cdot d)(b \cdot c),$$

and applying this identity to (12) yields

$$\hat{v} = \begin{pmatrix} (q_2 \wedge q_3) \cdot (m \wedge n) \\ (q_3 \wedge q_1) \cdot (m \wedge n) \\ (q_1 \wedge q_2) \cdot (m \wedge n) \end{pmatrix} = \begin{pmatrix} (q_2 \wedge q_3)^T \\ (q_3 \wedge q_1)^T \\ (q_1 \wedge q_2)^T \end{pmatrix} l.$$

In other words, we have just shown that the matrix $\mathcal{R}$ defined in Section 2 maps lines onto the direction of their image under $\mathcal{Q}$.

Applying this result to the lines λ and μ shows that the directions of the lines $\hat{\lambda}$ and $\hat{\mu}$ are given respectively by

$$\begin{cases} \hat{m}_1 \times \hat{m}_3 = \mathcal{R}\lambda, \\ \hat{m}_2 \times \hat{m}_3 = \mathcal{R}\mu, \end{cases}$$

and the lemma immediately follows.

Appendix B: Proof of Lemma 4

Here we establish the properties of the matrices $\mathcal{R}$ and $\mathcal{S}$. Let us define the column vectors of $\mathcal{R}^T$ as $R_1 = q_2 \wedge q_3$, $R_2 = q_3 \wedge q_1$, and $R_3 = q_1 \wedge q_2$.

These vectors are the Plücker coordinates of three lines R_1, R_2 and R_3 that intersect at the point of intersection of the three planes associated with the vectors q_1, q_2 and q_3. In particular we have the constraints

$$\begin{cases} (R_1|R_1) = 0, \\ (R_2|R_2) = 0, \\ (R_3|R_3) = 0, \end{cases} \text{ and } \begin{cases} (R_1|R_2) = 0, \\ (R_2|R_3) = 0, \\ (R_3|R_1) = 0. \end{cases}$$

Let us now turn our attention to S. We have

$$S = \begin{pmatrix} R_1 & R_2 & R_3 \end{pmatrix} \begin{pmatrix} R_1^T \\ R_2^T \\ R_3^T \end{pmatrix} = R_1 R_1^T + R_2 R_2^T + R_3 R_3^T.$$

In particular,

$$\begin{aligned} S_{16} - S_{25} + S_{34} &= (R_{11}R_{16} + R_{21}R_{26} + R_{31}R_{36}) - (R_{12}R_{15} + R_{22}R_{25} + R_{32}R_{35}) \\ &\quad + (R_{13}R_{14} + R_{23}R_{24} + R_{33}R_{34}) \\ &= (R_1|R_1) + (R_2|R_2) + (R_3|R_3) = 0. \end{aligned}$$

If we denote by S_1 to S_6 the columns of the matrix S, we have $S_i = R_{1i}R_1 + R_{2i}R_2 + R_{3i}R_3$. In particular, this means that the columns of S are the Plücker coordinate vectors of six lines, and these lines are all pairwise coplanar (in fact, they belong to the pencil generated by the lines R_1, R_2 and R_3). This yields 21 quadratic constraints of the form $(S_i|S_j) = 0$ ($i, j = 1, \ldots, 6$) on the entries of the matrix S. Note that these equations capture the linear dependency of the columns S_i and the fact that the matrix S has (at most) rank 3.

It is easily shown that only 12 of the quadratic constraints are linearly independent:

$$\begin{cases} (S_1|S_1) = 0, \\ (S_1|S_2) = 0, \\ (S_1|S_3) = 0, \\ (S_2|S_2) = 0, \\ (S_2|S_3) = 0, \\ (S_3|S_3) = 0, \end{cases} \quad \begin{cases} (S_4|S_1) = 0, \\ (S_4|S_2) = 0, \\ (S_4|S_3) = 0, \\ (S_5|S_1) = 0, \\ (S_5|S_2) = 0, \\ (S_6|S_1) = 0, \end{cases}$$

and that all other constraints are identical to one of these or its opposite (this is due to the symmetry to the matrix S).

It may also be of interest to note that the matrix $S = \mathcal{R}^T \mathcal{R}$ maps lines onto lines: the fact that the vector Sl verifies the Plücker constraint (4) for any Plücker vector l is easily verified analytically by using elementary properties of the cross product. If $\mathcal{P}$ denotes the 3×4 matrix formed by the top three rows of $\mathcal{Q}^T$, it is also interesting to note that $\mathcal{R}$ is the matrix called $\tilde{\mathcal{P}}$ by Faugeras and Papadopoulo [4], that maps lines in space onto the corresponding image lines under the perspective projection associated with the matrix $\mathcal{P}$. As shown by these authors, $\tilde{\mathcal{P}}^T$ maps points in the image plane onto the corresponding visual rays, yielding a different proof that S maps lines onto lines.

Appendix C: Proof of Property 1

We consider an arbitrary $n \times n$ symmetric matrix S with real coefficients, and diagonalize this matrix in an orthonormal basis as $S = \mathcal{U D U}^T$, where $\mathcal{D}$ is the diagonal matrix formed by the (possibly negative) eigenvalues of S and $\mathcal{U}$ is the orthogonal matrix formed by its eigenvectors. We seek the symmetric positive semidefinite (or *sps*) matrix $\hat{S}$ that minimizes $E^2 = |S - \hat{S}|_2^2$. Let us define $\hat{\mathcal{D}} = \mathcal{U}^T \hat{S} \mathcal{U}$, and note that $\hat{\mathcal{D}}$ is by construction positive semidefinite as well. Observing that $\hat{S} = \mathcal{U} \hat{\mathcal{D}} \mathcal{U}^T$, and using the invariance of the Frobenius form under orthogonal transformations reduces our original problem to minimizing $E^2 = |\mathcal{D} - \hat{\mathcal{D}}|_2^2$ over all sps matrices $\hat{\mathcal{D}}$.

Since $\mathcal{D}$ is diagonal, it is clear that among all matrices $\hat{\mathcal{D}}$ having the same diagonal (and in particular among all sps matrices having the same diagonal), the matrix minimizing E^2 must have zero off-diagonal entries. Our problem thus reduces to finding the sps diagonal matrix $\hat{\mathcal{D}}$ that minimizes

$$E^2 = \sum_{i=1}^{n}(D_i - \hat{D}_i)^2,$$

where D_i (resp. $\hat{D}_i$) denotes the ith diagonal entry of $\mathcal{D}$ (resp. $\hat{\mathcal{D}}$). The sps matrix $\hat{\mathcal{D}}$ has positive or zero diagonal elements. For entries $D_i \geq 0$, the value of $(D_i - \hat{D}_i)^2$ is clearly minimized by $\hat{D}_i = D_i$. On the other hand, when $D_i < 0$, $(D_i - \hat{D}_i)^2$ is clearly minimized by $\hat{D}_i = 0$. The result follows immediately.

References

1. B. Boufama, R. Mohr, and F. Veillon. Euclidian constraints for uncalibrated reconstruction. In *Proc. Int. Conf. Comp. Vision*, pages 466–470, Berlin, Germany, May 1993.
2. J. Costeira and T. Kanade. A multi-body factorization method for motion analysis. *Int. J. of Comp. Vision*, 29(3):159–180, September 1998.
3. O. Faugeras and B. Mourrain. On the geometry and algebra of the point and line correspondences between n images. Technical Report 2665, INRIA Sophia-Antipolis, 1995.
4. O. Faugeras and T. Papadopoulo. Gaussman-Cayley algebra for modeling systems of cameras and the algebraic equations of the manifold of trifocal tensors. Technical Report 3225, INRIA Sophia-Antipolis, 1997.
5. O.D. Faugeras. What can be seen in three dimensions with an uncalibrated stereo rig? In G. Sandini, editor, *Proc. European Conf. Comp. Vision*, volume 588 of *Lecture Notes in Computer Science*, pages 563–578, Santa Margherita, Italy, 1992. Springer-Verlag.
6. O.D. Faugeras. *Three-Dimensional Computer Vision*. MIT Press, 1993.
7. O.D. Faugeras. Stratification of 3D vision: projective, affine and metric representations. *J. Opt. Soc. Am. A*, 12(3):465–484, March 1995.
8. O.D. Faugeras, Q.-T. Luong, and S.J. Maybank. Camera self-calibration: theory and experiments. In G. Sandini, editor, *Proc. European Conf. Comp. Vision*, volume 588 of *Lecture Notes in Computer Science*, pages 321–334, Santa Margherita, Italy, 1992. Springer-Verlag.
9. A. Fitzgibbon and A. Zisserman. Automatic 3D model acquisition and generation of new images from video sequences. In *European Signal Processing Conference*, pages 311–326, Rhodes, Greece, 1998.
10. P.E. Gill and W. Murray. Newton-type methods for unconstrained and linearly constrained optimization. *Math. Programming*, 28:311–350, 1974.
11. C. Harris and M. Stephens. A combined edge and corner detector. In 4th *Alvey Vision Conference*, pages 189–192, Manchester, UK, 1988.
12. R. Hartley. Lines and points in three views and the trifocal tensor. *Int. J. of Comp. Vision*, 22(2):125–140, March 1997.
13. R.I. Hartley. An algorithm for self calibration from several views. In *Proc. IEEE Conf. Comp. Vision Patt. Recog.*, pages 908–912, Seattle, WA, June 1994.
14. R.I. Hartley. In defence of the 8-point algorithm. In *Proc. Int. Conf. Comp. Vision*, pages 1064–1070, Boston, MA, 1995.

15. R.I. Hartley, R. Gupta, and T. Chang. Stereo from uncalibrated cameras. In *Proc. IEEE Conf. Comp. Vision Patt. Recog.*, pages 761–764, Champaign, IL, 1992.
16. A. Heyden. *Geometry and algebra of multiple projective transformations*. PhD thesis, Lund University, Sweden, 1995.
17. A. Heyden and K. Åström. Minimal conditions on intrinsic parameters for Euclidean reconstruction. In *Asian Conference on Computer Vision*, Hong Kong, 1998.
18. A. Heyden and K. Åström. Flexible calibration: minimal cases for auto-calibration. In *Proc. Int. Conf. Comp. Vision*, pages 350–355, Kerkyra, Greece, September 1999.
19. J.J. Koenderink and A.J. Van Doorn. Affine structure from motion. *J. Opt. Soc. Am. A*, 8:377–385, 1990.
20. S.J. Maybank and O.D. Faugeras. A theory of self-calibration of a moving camera. *Int. J. of Comp. Vision*, 8(2):123–151, 1992.
21. R. Mohr, L. Quan, F. Veillon, and B. Boufama. Relative 3D reconstruction using multiple uncalibrated images. Technical Report RT 84-IMAG 12-LIFIA, LIFIA-IRIMAG, June 1992.
22. C.J. Poelman and T. Kanade. A paraperspective factorization method for shape and motion recovery. *IEEE Trans. Patt. Anal. Mach. Intell.*, 19(3):206–218, March 1997.
23. M. Pollefeys. *Self-calibration and metric 3D reconstruction from uncalibrated image sequences*. PhD thesis, Katholieke Universiteit Leuven, 1999.
24. M. Pollefeys, R. Koch, and L. Van Gool. Self-calibration and metric reconstruction in spite of varying and unknown internal camera parameters. *Int. J. of Comp. Vision*, 32(1):7–26, August 1999.
25. L. Quan. Self-calibration of an affine cameras from multiple views. *Int. J. of Comp. Vision*, 19:93–110, 1996.
26. R.B. Schnabel and E. Eskow. A new modified Cholesky factorization. *SIAM J. Sci. Comput.*, 11:1136–1158, 1990.
27. A. Shashua. Projective depth: a geometric invariant for 3D reconstruction from two perspective/orthographic views and for visual recognition. In *Proc. Int. Conf. Comp. Vision*, pages 583–590, Berlin, Germany, 1993.
28. A. Shashua. Trilinearity in visual recognition by alignment. In J.-O. Eklundh, editor, *Proc. European Conf. Comp. Vision*, volume 800 of *Lecture Notes in Computer Science*, pages 479–484. Springer-Verlag, 1994.
29. P. Sturm. Critical motion sequences for monocular self-calibration and uncalibrated Euclidean reconstruction. In *Proc. IEEE Conf. Comp. Vision Patt. Recog.*, pages 1100–1105, San Juan, Puerto Rico, June 1997.
30. C. Tomasi and T. Kanade. Shape and motion from image streams under orthography: a factorization method. *Int. J. of Comp. Vision*, 9(2):137–154, 1992.
31. P.H.S. Torr, A.W. Fitzgibbon, and A. Zisserman. The problem of degeneracy in structure and motion estimation from uncalibrated motion sequences. *Int. J. of Comp. Vision*, 32(1):27–44, August 1999.
32. R.Y. Tsai and T.S. Huang. Uniqueness and estimation of 3D motion parameters of rigid bodies with curved surfaces. *IEEE Trans. Patt. Anal. Mach. Intell.*, 6:13–27, 1984.
33. S. Ullman. *The Interpretation of Visual Motion*. The MIT Press, Cambridge, MA, 1979.
34. D. Weinshall and C. Tomasi. Linear and incremental acquisition of invariant shape models from image sequences. *IEEE Trans. Patt. Anal. Mach. Intell.*, 17(5), May 1995.
35. Z. Zhang, R. Deriche, O.D. Faugeras, and Q.-T. Luong. A robust technique for matching two uncalibrated images through the recovery of the unknown epipolar geometry. *Artificial Intelligence Journal*, 78:87–119, October 1995.

Discussion

1. **Rudolphe Mester, Frankfurt university**: I have two comments. The first comment is not only related to your work, but to a lot of other papers which have been presented. If you are talking about linear least squares, I think this is something that is relatively different from the normal usage of that term. What you have here is some kind of eigensystem problem as in Papadopoulo and Lourakis [1], not a linear equation system with errors. There are totally different mathematical methods used to describe perturbations of such systems.
Secondly I refer to those normalizations that you need in order to consider the statistical structure of the errors in the input data, which might be very significant. These can be performed using some rather well known techniques from numerical linear algebra, such as equilibration, where the normalization techniques proposed by Richard Hartley in 1995 and other proposals are just special cases. So, partially at least, I think there are techniques available to improve the robustness of your method against these errors.
Jean Ponce: I know that these methods exist and I did some work in the past with Peter Meer and used some of his techniques. We did it for bilinear systems where it worked very well. But for more complex systems like this one that may not be the case.
Bill Triggs, INRIA Rhône-Alpes: Just a comment. Normalization and total least squares work well for some problems, but for multiresultant style polynomial solvers we found that total least squares reweighting (pre- and post-multiplying the multiresultant matrix with weighting matrices) made essentially no difference. The problem is that the errors come from the polynomial coefficients, which are repeated many times in the multiresultant matrix in a patterned structure. So the matrix coefficient error model is sparse and very highly structured and correlated, and it seems to be poorly approximated by the left-and-right-rescaled-Frobenius-norm error model that total least squares normalization assumes. Jean's technique also involves quasi-linearization of a polynomial system, so it is likely to have similar problems.

References

1. T. Papadopoulo and M. Lourakis. Estimating the jacobian of the singular value decomposition: Theory and applications. In *Proc. European Conference on Computer Vision*, pages 554–570, 2000.

A Progressive Scheme for Stereo Matching

Zhengyou Zhang and Ying Shan

Microsoft Research, One Microsoft Way, Redmond, WA 98052, USA
zhang@microsoft.com,
WWW home page: http://research.microsoft.com/~zhang/

Abstract. Brute-force dense matching is usually not satisfactory because the same search range is used for the entire image, yielding potentially many false matches. In this paper, we propose a progressive scheme for stereo matching which uses two fundamental concepts: the disparity gradient limit principle and the least commitment strategy. The first states that the disparity should vary smoothly almost everywhere, and the disparity gradient should not exceed a certain limit. The second states that we should first select only the most reliable matches and therefore postpone unreliable decisions until enough confidence is accumulated. Our technique starts with a few reliable point matches obtained automatically via feature correspondence or through user input. New matches are progressively added during an iterative matching process. At each stage, the current reliable matches constrain the search range for their neighbors according to the disparity gradient limit, thereby reducing potential matching ambiguities of those neighbors. Only unambiguous matches are selected and added to the set of reliable matches in accordance with the least commitment strategy. In addition, a correlation match measure that allows rotation of the match template is used to provide a more robust estimate. The entire process is cast within a Bayesian inference framework. Experimental results illustrate the robustness of our proposed dense stereo matching approach.

Keywords: Stereo vision, Stereo matching, Disparity gradient limit, Least commitment, Progressive matching, Bayesian inference, Correlation, Image registration.

1 Introduction

Over the years numerous algorithms for image matching have been proposed. They can roughly be classified into two categories:

Feature matching. They first extract salient primitives from the images, such as corners and edge segments, and match them across two or more views. An image can then be described by a graph with primitives defining the nodes and geometric relations defining the links. Matching becomes finding the mapping of graphs: subgraph isomorphism. Some heuristics such as assuming affine transformation between images are usually introduced to reduce the complexity. These methods are fast because only a small subset of the image pixels are used, but may fail if the chosen primitives cannot be reliably detected in the images. They only produce a very coarse 3D model of the actual scene. The following list of references is by no means exhaustive: [9,13,15,1,5]

Template matching. They attempt to correlate image patches across views, assuming that they present some similarity [8,10,7,18,20]. The underlying assumption appears to be a valid one for relatively textured areas and for image pairs with small difference; however it may be wrong at occlusion boundaries and within featureless regions. Although these algorithms produce a dense 3D reconstruction of the actual scene, brute-force matching is usually not satisfying because of potentially many false matches.

All above stereo matching algorithms suffer from the difficulty in specifying an appropriate search range and the inability to adapt the search range depending on the observed scene structure.

In this paper, we propose a progressive scheme that, to some extent, combines these two approaches. It starts with a few reliable point matches obtained automatically via feature correspondence or through user input. It then tries to find progressively more pixel matches based on two fundamental concepts: disparity gradient limit principle and least commitment strategy. The disparity gradient limit principle states that the disparity should vary smoothly almost everywhere, and the disparity gradient should not exceed a certain value. This defines the search range for candidate matches. The least commitment strategy states that we should first select only the most reliable matches and therefore postpone an unreliable decision until enough confidence is accumulated. New matches are progressively added during an iterative matching process. At each stage, the current reliable matches constrain the search range for their neighbors according to the disparity gradient limit, thereby reducing potential matching ambiguities of those neighbors. Only unambiguous matches are selected and added to the set of reliable matches in accordance with the least commitment strategy.

Lhuillier and Quan recently reported a matching algorithm using a similar idea [11]. They also start with a few reliable point matches, but the technique to find more matches is very different from ours. They first choose the best match, and look for additional matches in their 5×5 neighborhood. Therefore, they only consider one match each time and propagate it in a very small area, while we consider all current matches simultaneously and do not restrict the propagation within a very small area. Chen and Medioni [3] uses a very similar strategy to that of Lhuillier and Quan, but work with a volumetric representation.

The paper is organized as follows. Section 2 presents the disparity gradient limit principle and the least commitment strategy, and introduces a scheme for progressive matching. Section 3 describes the implementation details on how disparities are predicted and estimated, which is formulated within a Bayesian inference framework. Section 4 proposes a new correlation technique designed for cameras in general position. Section 5 provides experimental results, including intermediate ones, with two sets of real data. Section 6 concludes the paper with a discussion on future work.

2 A Progressive Scheme

We first describe the two fundamental concepts, namely the disparity gradient limit principle and the least commitment strategy. We then present a simple progressive scheme which starts a few seed matches and then tries to find progressively more pixel matches based on these two concepts.

2.1 Disparity Gradient Limit Principle

Disparity is directly related to depth. Disparity changes coincide with depth changes. The disparity gradient limit principle states that the disparity should vary smoothly almost everywhere, and the disparity gradient should not exceed a certain value. Psychophysical studies have provided evidence that in order for the human visual system to binocularly fuse two dots of a simple stereogram, the disparity gradient (ratio of the disparity difference) between the dots to their cyclopean separation must not exceed a limit of 1 [2,16]. Objects in the world are usually bounded by continuous opaque surfaces, and disparity gradient can be considered as a simple measure of continuity. The disparity gradient limit principle provides a constraint on scene jaggedness embracing simultaneously the ideas of opacity, scene continuity, and continuity between views [14]. It has been used in several successful stereo matching algorithms including the PMF algorithm [15] to resolve matching ambiguity.

The disparity gradient limit principle is used differently in our work, as we will explain in details in Section 3.1. It is exploited to estimate the uncertainty of the predicted disparity for a particular pixel, and the uncertainty is then used to define the search ranges for candidate matches.

2.2 Least Commitment Strategy

The least commitment strategy states that we should first select only the most reliable decisions and therefore postpone an unreliable decision until enough confidence is accumulated. It is a powerful strategy used in Artificial Intelligence, especially in action planning [22,19]. Since no irreversible decision is made (i.e. all decisions made are reliable), this principle offers significant flexibility in avoiding locking search into a possibly incorrect step where an expensive refinement such as backtracking has to be exploited.

The least commitment strategy is explored in our algorithm in four ways (abbreviated as STAB):

Search range. Matching criterion such as correlation is local and heuristic. If the match of a pixel has to be searched in a wide range, there is a high probability that the found match is not a correct one. It is preferable to defer matching of these pixels as late as possible because the search range may be reduced later after more reliable matches are established.

Texture. A pixel is more discriminating in a highly textured neighborhood than others. It is difficult to distinguish pixels in the same neighborhood having similar intensity. Therefore, we can expect to have more reliable matches for pixels in areas with strong textures, and thus try to match them first.

Ambiguity. We may find several candidate matches for a pixel. Rather than using expensive techniques such as dynamic programming to resolve the ambiguity, we simply defer the decision. Once more reliable matches are found in the future, the ambiguity will become lower because of a better disparity estimate with smaller uncertainty.

Bookkeeping. If a pixel does not have any candidate match, it is probably occluded by others or is not in the field of view of the other camera, then we do not need to search for its match in the future. Similar, if a pixel has already found a match, further search is not necessary. We bookkeep both types of pixels for efficiency.

2.3 A Progressive Stereo Matching Algorithm

We can now outline the proposed progressive algorithm. Details will be given in the following sections.

A pixel in the first image has three labels: MATCHED (already matched), NOMATCH (no candidate matches found), and UNKNOWN (not yet decided). All pixels are initially labeled as UNKNOWN.

For a pixel which is labeled UNKNOWN, we compute a list of candidate pixels in the second image which satisfy the epipolar constraint and disparity gradient limit constraint. We use the normalized cross correlation as our matching criterion. For a pair of pixels between two images, we compute the normalized cross correlation score between two small windows, called *correlation windows*, centered at the pixels. The correlation score ranges from -1, for two correlation windows which are not similar at all, to $+1$, for two correlation windows which are identical. The pair of pixels are considered as a potential match if the correlation score is larger than a predefined threshold T_C. The list of candidate pixels are ordered on the epipolar line, and the correlation scores form a curve. If there is only one peak on the correlation curve exceeding the threshold T_C, then the pixel at the peak is considered as the match of the given pixel in the first image, and the given pixel is labeled as MATCHED. If there is no peak exceeding the threshold T_C, we label the given pixel as NOMATCH, as we mentioned earlier. If there are two or more peaks exceeding T_C, the matching is ambiguous, and according to the least commitment principle, we simply leave it as is. We iterate this procedure until no more matches can be found or the maximum number of iteration is attained.

As we described earlier, pixels in highly textured areas are considered first. Textureness is measured as the sample deviation of the intensity within a correlation window. In order for a pixel in the first image to be considered, its sample deviation must be larger than a threshold T_{σ_I}. The threshold T_{σ_I} evolves with iteration. It is given by a monotonic function *ThresholdSigmaIntensity* which never increases with iteration.

Similarly, if a given pixel in the first image has a large uncertainty of its disparity vector, this pixel should be considered as late as possible. In order for a pixel to be considered, the standard deviation of its predicted disparity vector must be smaller than a threshold T_{σ_D}. The threshold T_{σ_D} evolves with iteration. It is given by a monotonic function *ThresholdSigmaDisparity* which never decreases with iteration. That is, we, at the beginning, only considered pixels that have a good prediction of the disparity vector.

Please note that the above description is outlined only to present the essential ideas. The actual implementation of several components such as correlation computation is different, as we will describe in the next section.

The pseudo C++ code of the algorithm is summarized in Figure 1.

```
iteration = 0;
while (the maximum number of iterations is not reached)
    and (more matches are found) {
    T_σI = ThresholdSigmaIntensity(iteration);
    T_σD = ThresholdSigmaDisparity(iteration);
    for (every pixel labeled UNKNOWN in the first image) {
        estimate the disparity vector and its uncertainty;
        if (σI_of_the_pixel < T_σI)
            continue;   // not enough textured
        if (σD_of_the_pixel < T_σD)
            continue;   // too much uncertainty for its match
        compute the list of candidate pixels in the second image;
        compute the correlation score C for each candidate pixel;
        if (there is one peak on the correlation curve)
            and (its C > TC) {
            update its disparity vector;
            label the pixel as MATCHED.
        }
        else if (there is no candidate whose C > TC) {
            label the pixel as NOMATCH.
        }
    }
}
```

Fig. 1. Pseudo C++ code of the progressive stereo matching algorithm.

The above algorithm has a number of important properties:

Progressiveness. Because of bookkeeping, the number of pixels examined in each iteration becomes smaller. Also, as we will show later, the search range for a pixel is reduced when we update the disparity with more matched pixels. This property guarantees that the iterative procedure is actually making some progress and that the search space is being reduced.

Monotonicity. Because of the monotonicity of functions *ThresholdSigmaIntensity* and *ThresholdSigmaDisparity*, threshold T_{σ_I} is getting smaller and threshold T_{σ_D} is getting larger with the progress of the algorithm. This means that the probability that a pixel labeled as UNKNOWN is selected for matching test becomes higher, eventually resulting more MATCHED/NOMATCH pixels. Together with the update of disparity vectors and their uncertainty, this property guarantees that the set of UNKNOWN pixels considered is truly different from that prior to refinement, "different" in the sense of the actual pixels considered and also of their candidate pixels to match in the other image.

Completeness. This property says that adding more MATCHED/NOMATCH pixels will not lose any potential matches. This is desirable because it means that an expensive refinement such as backtracking is never performed. The above proposed algorithm clearly satisfies this property because of the least commitment strategy, provided that the disparity gradient limit constraint is satisfied over the entire observed scene.

The completeness property of our algorithm does not imply that as the final result each pixel must be labeled either MATCHED or NOMATCH. Indeed, pixels within a uniform color region may still be labeled as UNKNOWN. However, from the neighboring matched pixels, these pixels have an estimate of their disparity vectors that can be used if necessary, for example, for image-based rendering.

3 Implementation Details

In this section, we provide the details in implementing the progressive algorithm described in the last section. Basically, for each pixel labeled UNKNOWN, we need to do two things: prediction the disparity and its uncertainty, based on the information provided by the neighboring matched pixels; estimation of its disparity based on the information contained in the images.

If we formulate the problem in terms of Bayesian inference (see e.g. [21]), the first corresponds to the prior density distribution of the disparity, $p(d|\mathbf{m}, B)$, where d is the disparity of the given pixel $\mathbf{m}$, and B denote the relevant background information at hand such as the epipolar geometry and the set of already matched pixels. The second corresponds to the sampling distribution $p(I'|d, \mathbf{m}, B)$, or the likelihood of the observed data (i.e., the second image I') given d, $\mathbf{m}$ and B. Bayes' rule can then be used to combine the information in the data with the prior probability, which yields the posterior density distribution

$$p(d|I', \mathbf{m}, B) = \frac{p(I'|d, \mathbf{m}, B)p(d|\mathbf{m}, B)}{p(I'|\mathbf{m}, B)}, \qquad (1)$$

where $p(I'|\mathbf{m}, B)$ does not depend on d and can be considered as a constant because the second image I' is fixed. We can thus omit the factor $p(I'|\mathbf{m}, B)$ and work on the unnormalized posterior density distribution $p(I'|d, \mathbf{m}, B)p(d|\mathbf{m}, B)$, still denoted by $p(d|I', \mathbf{m}, B)$ to abuse the notation. Appropriate computations to summarize $p(d|I', \mathbf{m}, B)$ are finally performed in order to decide whether the pixel under consideration should be labeled MATCHED or NOMATCH, or kept as UNKNOWN for future decision.

3.1 Prediction of the Disparity and Its Uncertainty

Before introducing our work, it is helpful to define disparity and disparity gradient and summarize the related results obtained by others.

Disparity is well defined for parallel cameras (i.e., the two image planes are the same) [6]. Without loss of generality, the horizontal axis is assumed to be aligned in both images. Given a pixel of coordinates (u, v) in the first image and its corresponding pixel of coordinates (u', v') in the second image, disparity is defined as the difference $d = v' - v$. Disparity is inversely proportional to the distance of the 3D point to the cameras. A disparity of 0 implies that the 3D point is at infinity.

Consider now two 3D points whose projections are $\mathbf{m}_1 = [u_1, v_1]^T$ and $\mathbf{m}_2 = [u_2, v_2]^T$ in the first image, and $\mathbf{m}'_1 = [u'_1, v'_1]^T$ and $\mathbf{m}'_2 = [u'_2, v'_2]^T$ in the second image ($u'_1 = u_1$ and $u'_2 = u_2$ in the parallel cameras case). Their disparity gradient is defined to be the ratio of their difference in disparity to their distance in the cyclopean image.[1] In the first image, the disparity gradient is given by

$$DG = \left| \frac{d_2 - d_1}{v_2 - v_1 + (d_2 - d_1)/2} \right| . \qquad (2)$$

Experiments in psychophysics have provided evidence that human perception imposes the constraint that the disparity gradient DG is upper-bounded by a limit K. That is, if a point on an object is perceived, neighboring points having $DG > K$ are simply not perceived correctly. The limit $K = 1$ was reported in [2]. The theoretical limit for opaque surfaces is $K = 2$ to ensure that the surfaces are visible to both eyes [14]. Although the range of allowable surfaces is large with $K = 2$, disambiguating power is weak because false matches receive and exchange as much support as correct ones. Another extreme limit is $K \approx 0$, which allows only nearly front-parallel surfaces, and this has been used locally in the stereogram matching algorithm described in [12]. In the PMF algorithm, the disparity gradient limit K is a free parameter, which can be varied over range $(0, 2)$. An intermediate value, e.g., between 0.5 and 1, allow selection of a convenient trade-off point between allowable scene surface jaggedness and disambiguating power because it turns out that most false matches produce relatively high disparity gradients [14]. Again, as reported in [14], less than 10% of world surfaces viewed at more than 26cm with 6.5cm of eye separation will present with disparity gradient larger than 0.5. This justifies use of a disparity gradient limit well below the theoretical value (of 2) without imposing strong restrictions on the world surfaces that can be fused by the stereo algorithm.

When the cameras are in general position, it is not reasonable to hope to define a scalar disparity as a simple function of the image coordinates of two pixels in correspondence [6]. In this work, we simply use a vector $\mathbf{d} = [u' - u, v' - v]^T$, called the disparity vector. This is the same as the flow vector used in optical flow computation. If a scalar value is necessary, we use $d = \|\mathbf{d}\|$ and call it the disparity. If we look at objects that are smooth almost everywhere, both $\mathbf{d}$ and d should vary smoothly. Similar to (2), for two points $\mathbf{m}_1$ and $\mathbf{m}_2$

---

[1] For a pair of pixels in correspondence with coordinates (u, v) and (u', v'), the cyclopean image point is at $((u + u')/2, (v + v')/2)$

in the first image, we define the disparity gradient as

$$DG = \frac{\|\mathbf{d}_2 - \mathbf{d}_1\|}{\|\mathbf{m}_2 - \mathbf{m}_1 + (\mathbf{d}_2 - \mathbf{d}_1)/2\|} . \quad (3)$$

Imposing the gradient limit constraint $DG \leq K$, we have

$$\|\mathbf{d}_2 - \mathbf{d}_1\| \leq K\|\mathbf{m}_2 - \mathbf{m}_1 + (\mathbf{d}_2 - \mathbf{d}_1)/2\| .$$

Using inequality $\|\mathbf{v}_1 + \mathbf{v}_2\| \leq \|\mathbf{v}_1\| + \|\mathbf{v}_2\|$ for any vectors $\mathbf{v}_1$ and $\mathbf{v}_2$, we obtain

$$\|\mathbf{d}_2 - \mathbf{d}_1\| \leq K\|\mathbf{m}_2 - \mathbf{m}_1\| + K\|(\mathbf{d}_2 - \mathbf{d}_1)/2\|$$

which leads immediately, for $K < 2$, to

$$\|\mathbf{d}_2 - \mathbf{d}_1\| \leq \frac{2K}{2-K}D , \quad (4)$$

where $D = \|\mathbf{m}_2 - \mathbf{m}_1\|$ is the distance between $\mathbf{m}_1$ and $\mathbf{m}_2$. We immediately have the following result:

Lemma 1. *Given a pair of matched points $(\mathbf{m}_1, \mathbf{m}'_1)$ and a point $\mathbf{m}_2$ in the neighborhood of $\mathbf{m}_1$, the corresponding point $\mathbf{m}'_2$ that satisfies the disparity gradient constraint with limit K must be inside a disk centered at $\mathbf{m}_2 + \mathbf{d}_1$ with radius equal to $\frac{2K}{2-K}D$, which we call the* continuity disk.

In other words, in absence of other knowledge, the best prediction of the disparity of $\mathbf{m}_2$ is equal to $\mathbf{d}_1$ with the continuity disk defining its uncertainty.

We may want to favorite the actual disparity to be at the central part of the continuity disk. We may also want to consider a small probability that the actual disparity is outside of the continuity disk, due to occlusion or surface discontinuity. We therefore model the uncertainty as an isotropic Gaussian distribution with standard deviation equal to half of the radius of the continuity disk. More precisely, given a pair of matched points $(\mathbf{m}_i, \mathbf{m}'_i)$, the disparity of a point $\mathbf{m}$ is modeled as

$$\mathbf{d} = \mathbf{d}_i + D_i \mathbf{n}_i , \quad (5)$$

where $\mathbf{d}_i = \mathbf{m}'_i - \mathbf{m}_i$, $D_i = \|\mathbf{m} - \mathbf{m}_i\|$, and $\mathbf{n}_i \sim N(\mathbf{0}, \sigma_i^2 \mathbf{I})$ with $\sigma_i = K/(2-K)$. Note that disparity $\mathbf{d}_i$ also has its own uncertainty due to limited image resolution. The density distribution of $\mathbf{d}_i$ is also modeled in our work as a Gaussian, i.e., $p(\mathbf{d}_i) = N(\mathbf{d}_i | \bar{\mathbf{d}}_i, \sigma_{d_i}^2 \mathbf{I})$. It follows that the density distribution of disparity $\mathbf{d}$ is given by

$$p(\mathbf{d}|(\mathbf{m}_i, \mathbf{m}'_i), \mathbf{m}) = N(\mathbf{d}|\bar{\mathbf{d}}_i, (\sigma_{d_i}^2 + D_i^2 \sigma_i^2)\mathbf{I}) . \quad (6)$$

If we are given a set of point matches $\{(\mathbf{m}_i, \mathbf{m}'_i) | i = 1, \ldots, n\}$, we then have n independent predictions of disparity $\mathbf{d}$ as given by (6). The prior density distribution of the disparity, $p(\mathbf{d}|\mathbf{m}, B)$, can be obtained by combining these predictions with the minimum variance estimator, i.e.,

$$p(\mathbf{d}|\mathbf{m}, B) = N(\mathbf{d}|\bar{\mathbf{d}}, \sigma^2 \mathbf{I}) , \quad (7)$$

where

$$\bar{\mathbf{d}} = \Big(\sum_{i=1}^{n} \frac{1}{\sigma_{d_i}^2 + D_i^2 \sigma_i^2}\Big)^{-1} \sum_{i=1}^{n} \frac{1}{\sigma_{d_i}^2 + D_i^2 \sigma_i^2} \bar{\mathbf{d}}_i$$

$$\sigma^2 = \Big(\sum_{i=1}^{n} \frac{1}{\sigma_{d_i}^2 + D_i^2 \sigma_i^2}\Big)^{-1}.$$

A more robust version is first to identify the Gaussian with smallest variance, and then to combine it with those Gaussians whose means fall within two or three standard deviations.

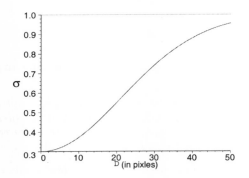

Fig. 2. Function of σ_i (related to the disparity gradient limit) w.r.t. the distance to a matched pixel. See (8).

It remains the problem of choosing σ_i, which as mentioned earlier is related to the disparity gradient limit K. In the PMF algorithm, K is set to a value between 0.5 and 1, which is equivalent to a value between $1/3$ and $1/2$ for our σ_i. Considering that the disparity gradient constraint is still a local one, it should become less restrictive when the point being considered is away from a matched point. Hence, we specify a range $[\sigma_{\min}, \sigma_{\max}]$, and σ_i is given by

$$\sigma_i = (\sigma_{\max} - \sigma_{\min})(1 - \exp(-D_i^2/\tau^2)) + \sigma_{\min}. \quad (8)$$

When $D_i = 0$, $\sigma_i = \sigma_{\min}$; when $D_i = \infty$, $\sigma_i = \sigma_{\max}$. The parameter τ controls how fast the transition from $\sigma_{\min}$ to $\sigma_{\max}$ is expected. In our implementation, $\sigma_{\min} = 0.3$ pixels, $\sigma_{\max} = 1.0$ pixel, and $\tau = 30$. This is equivalent to $K_{\min} = 0.52$ and $K_{\max} = 1.34$. Figure 2 displays how σ_i varies with respect to the distance D_i. From many images we have tried, this strategy works well.

3.2 Computation of the Disparity Likelihood

We now proceed to compute the sampling distribution $p(I'|d, \mathbf{m}, B)$, or the likelihood of the observed data (i.e., the second image I') given d, $\mathbf{m}$ and B.

Because of the epipolar constraint, we do not need to compute the density for each pixel in I'. Furthermore, we do not even need to compute the density for each pixel on the epipolar line of $\mathbf{m}$ because of the prior density computed in (7). The list of pixels of interest, called the *candidate pixels* and denoted by $Q(\mathbf{m})$, is the intersection of the epipolar line of $\mathbf{m}$ with the continuity disk defined in Lemma 1.

The densities are related to the correlation scores C_j between $\mathbf{m}$ in the first image and each candidate pixel $\mathbf{m}'_j \in Q(\mathbf{m})$ in the second image. Instead of using the standard correlation technique based on two rectangular windows, we have developed a new one which is well adapted for two images in general position. We defer its presentation to Section 4. For the moment, it suffices to say that the correlation score C is between -1 (when they are not similar at all) and $+1$ (when they are identical). Finally, correlation scores are mapped to densities by adding 1 followed by a normalization. More precisely, the correlation score C_j of a pixel $\mathbf{m}'_j$ is converted into a density as

$$p(I'(\mathbf{m}'_j)|\mathbf{d}^{(j)}, \mathbf{m}, B) = \frac{C_j + 1}{\sum_{k \in Q(\mathbf{m})}(C_k + 1)}, \qquad (9)$$

where $\mathbf{d}^{(j)} = \mathbf{m}'_j - \mathbf{m}$.

3.3 Inference from the Posterior Density

The posterior density distribution $p(d|I', \mathbf{m}, B)$ is simply multiplication of $p(I'(\mathbf{m}'_j)|\mathbf{d}^{(j)}, \mathbf{m}, B)$ in (9) with $p(\mathbf{d}^{(j)}|\mathbf{m}, B)$ in (7) for each candidate pixel $\mathbf{m}'_j$.

Based on $p(d|I', \mathbf{m}, B)$, we can do a number of things. If there is only one prominent peak, the probability that this is a correct match is very high, and we thus make the decision and label the pixel in the first image MATCHED. If there are two or more prominent peak, the matching ambiguity is high, i.e., the probability of making a wrong decision is high. Following the least commitment principle, we leave this pixel to evolve. If there is no prominent peak at all, the probability that the corresponding point in the second image is not visible is very high (either occluded by others or out of the field of view), and we label the pixel in the first image NOMATCH.

In order to facilitate the task of choosing an appropriate threshold on the posterior density distribution, and since anyway we are working with the *unnormalized* posterior density distribution, we normalize the prior and likelihood functions differently. The prior in (7) is multiplied by $\sigma\sqrt{2\pi}$ so that the maximum is equal to one. The likelihood in (9) is changed to $(C_j + 1)/2$ so that it is equal to 1 for identical pixels and 0 for completely different pixels. A peak in the posterior density distribution is considered as a prominent one if its value is larger than 0.3, which corresponds to, e.g., the situation where $C_j = 0.866$ and the disparity is at 1.5σ.

4 A New Correlation Technique

The correlation technique described in this section is designed for stereo cameras in general position.

Table 1. Number of matched pixels in each iteration

iteration	0	1	2	3	4	5	6
T_{σ_I}		7	6	5	4	3	2
T_{σ_D}		12	14	16	18	20	20
Books	141	455	712	939	1239	1440	1500
NMars	153	421	1249	2036	2360	2651	2741

Consider a pair of points **m** and **m**′ as shown in Fig. 3, where the corresponding epipolar lines l and l' are also drawn. We can easily compute a Euclidean transformation

$$\mathbf{m}'_i = \mathbf{R}(\theta)(\mathbf{m}_i - \mathbf{m}) + \mathbf{m} , \tag{10}$$

where $\mathbf{R}(\theta)$ is a 2D rotation matrix with rotation angle equal to θ, the angle between the two epipolar lines. It sends **m** to **m**′ and a point on l to a point on l'.

Choose a rectangular window centered at **m** with one side parallel to the epipolar line. A point $\mathbf{m}_i$ corresponds to a point $\mathbf{m}'_i$ given by (10). Point $\mathbf{m}'_i$ is usually not on the pixel grid, and its intensity is computed through bilinear interpolation from its four neighboring pixels. Correlation score is then computed between points $\mathbf{m}_i$ in the correlation window and points $\mathbf{m}'_i$ according to (10). We use the normalized cross correlation [6] which is equal to 1 for two identical sets of pixels and -1 for two completely different sets.

If two epipolar lines are both horizontal or vertical, the new technique will be equivalent to the standard one.

An even more elaborate way to compute the correlation is to weight differently each point: Pixels in the central part have more weights than those near the border. In our implementation, the size of correlation window is 11 pixels along the epipolar line and 9 pixels in the other direction. The pixels are weighted by a 2D Gaussian with standard deviation equal to 11 pixels along the epipolar line and 9 pixels in the other direction.

5 Experimental Results

We have conducted experiments with several sets of real data, and very promising results have been obtained. In this section, we report two of them: one is an office scene with books, called Scene **Books** (see Fig. 4); another is a scene with rocks

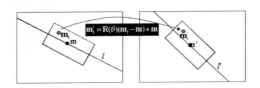

Fig. 3. The new correlation technique for stereo cameras in general position.

Fig. 4. Scene Books: Initial point matches indicated by the disparity vectors together with the Delaunay triangulation in the first image.

from INRIA, call Scene NMars (see Fig. 9). Although the images in Scene Books are color, only black/white information is used. The image resolution is 740×480 for Scene Books, and 512×512 for Scene NMars.

To reduce computation cost, instead of using all previously found matches in predicting disparities and their uncertainties, we only use three neighboring points defined by the Delaunay triangulation [17]. The dynamic Delaunay triangulation algorithm described in [4] is used because of its efficiency in updating the triangulation when more point matches are available. It is reasonable to use only three neighboring points because other points are usually much farther away, resulting in a larger uncertainty in its predicted disparity, hence contributing little to the combined prediction of the disparity given in (7).

The initial set of point matches, together with the fundamental matrix, were obtained automatically using the robust image matching technique described in [23]. All parameters are the same for both data sets. The search range was $[-60, 60]$ (pixels) for both horizontal and vertical directions.

All parameters in our algorithm are the same for both data sets. In particular, the values of functions $ThresholdSigmaIntensity$ and $ThresholdSigmaDisparity$ with respect to the iteration number are given in the second and third rows of Table 1. For example, for iteration 4, $T_{\sigma_I} = 4$ and $T_{\sigma_I} = 18$. In Table 1, we also provide the number of matches after each iteration. The number of matches for iteration 0 indicates the number of initial matches found by the robust matching algorithm. Note that instead of working on each pixel, we actually consider only one every four pixels because of the memory limitation in our Delaunay triangulation algorithm.

The initial set of point matches for Scene Books is shown in Fig. 4. Based on these, the disparity and its uncertainty were predicted, which are shown in Fig. 5. On the left, the disparity vectors are displayed for every 10 pixels and their lengths are half of their actual magnitudes. On the right, the standard deviation of the predicted disparities is shown in gray levels after having multiplied by 5 and truncated at 255. Therefore, "black" pixels in that image mean that the predicted disparities are quite reliable, while "white" pixels implies that the predicted disparities are very uncertain. The intermediate results after iteration

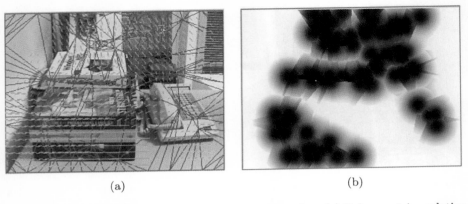

Fig. 5. Scene Books: Results with the initial point matches. (a) Delaunay triangulation and the predicted disparity vectors; (b) Predicted deviation of the disparity vectors.

Fig. 6. Scene Books: Results after the second iteration. (a) Delaunay triangulation and the predicted disparity vectors; (b) Predicted deviation of the disparity vectors.

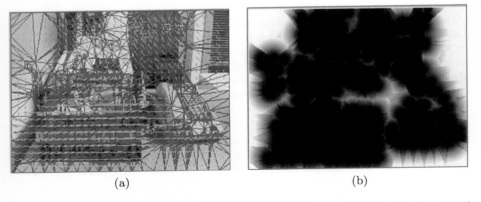

Fig. 7. Scene Books: Results after the sixth iteration. (a) Delaunay triangulation and the predicted disparity vectors; (b) Predicted deviation of the disparity vectors.

Fig. 8. Scene Books: Views of the 3D reconstruction with texture mapped from the 1st image.

Fig. 9. Scene NMars: Initial point matches indicated by the disparity vectors together with the Delaunay triangulation in the first image.

2, and 6 are shown in Fig. 6, and Fig. 7. We can observe clearly the fast evolution of the matching result. The uncertainty image becomes darker quickly. As we know the intrinsic parameters of the camera with which the images were taken, 3D Euclidean reconstruction can be obtained, two views of which are shown in Fig. 8. We can see that the book structure has been precisely recovered.

Similar results have been obtained with Scene NMars. As can be observed from Fig. 9, the lower part of the scene cannot be matched because the disparity is larger than the prefixed range (plus/minus a quarter of the image width). The predicted disparity vectors and their uncertainty computed from the initial set of matches are shown in Fig. 10, while those after iteration 6 are shown Fig. 12. It is clear that our progressive stereo algorithm is capable of finding matches with large disparity, the lower part of the scene in our case, even if the initial search range is large enough. 3D Euclidean reconstruction was also computed, two views of which are shown in Fig. 13.

 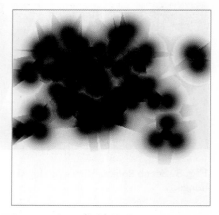

Fig. 10. Scene NMars: Results with the initial point matches. (left) Delaunay triangulation and the predicted disparity vectors; (right) Predicted deviation of the disparity vectors.

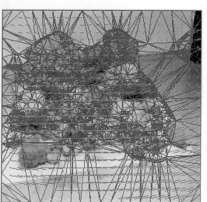 

Fig. 11. Scene NMars: Results after the third iteration.

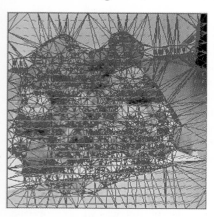 

Fig. 12. Scene NMars: Results after the sixth iteration.

Fig. 13. Scene NMars: Views of the 3D reconstruction with texture mapped from the 1st image.

6 Conclusions

In this paper, we have proposed a progressive scheme for stereo matching. It starts with a few reliable point matches obtained either manually from user input or automatically with feature-based stereo matching. It then tries to find progressively more pixel matches based on two fundamental concepts: disparity gradient limit principle and least commitment strategy. Experimental results have proven the robustness of our proposed dense stereo matching approach.

We have also cast the disparity estimation in the framework of Bayesian inference, and have developed a new correlation technique well adapted for cameras in general position.

There are a number of ways to extend the current algorithm. For example, the current implementation only estimate disparities with pixel precision. One of our future work consists in produce disparities with subpixel precision. We will also investigate in an even more efficient implementation.

References

1. N. Ayache and B. Faverjon. Efficient registration of stereo images by matching graph descriptions of edge segments. *The International Journal of Computer Vision*, 1(2), April 1987.
2. P. Burt and B. Julesz. A gradient limit for binocular fusion. *Science*, 208:615–617, 1980.
3. Q. Chen and G. Medioni. A volumetric stereo matching method: Application to image-based modeling. In *Proceedings of the IEEE Conference on Computer Vision and Pattern Recognition*, volume 1, pages 29–34, Colorado, June 1999. IEEE Computer Society.
4. O. Devillers, S. Meiser, and M. Teillaud. Fully dynamic Delaunay triangulation in logarithmic expected time per operation. *Comput. Geom. Theory Appl.*, 2(2):55–80, 1992.
5. Umesh R. Dhond and J.K. Aggarwal. Structure from stereo - a review. *IEEE Transactions on Systems, Man, and Cybernetics*, 19(6):1489–1510, 1989.

6. Olivier Faugeras. *Three-Dimensional Computer Vision: a Geometric Viewpoint.* MIT Press, 1993.
7. Pascal Fua. A parallel stereo algorithm that produces dense depth maps and preserves image features. *Machine Vision and Applications*, 6(1):35–49, Winter 1993. Available as INRIA research report 1369.
8. A. Goshtasby, S. H. Gage, and J. F. Bartholic. A two-stage cross correlation approach to template matching. *IEEE Transactions on Pattern Analysis and Machine Intelligence*, 6(3):374–378, May 1984.
9. W.E.L. Grimson. Computational experiments with a feature based stereo algorithm. *IEEE Transactions on Pattern Analysis and Machine Intelligence*, 7(1):17–34, 1985.
10. M.J. Hannah. A system for digital stereo image matching. *Photogrammetric Engeneering and Remote Sensing*, 55(12):1765–1770, December 1989.
11. M. Lhuillier and L. Quan. Image interpolation by joint view triangulation. In *Proceedings of the IEEE Conference on Computer Vision and Pattern Recognition*, volume 2, pages 139–145, Colorado, June 1999. IEEE Computer Society.
12. D. Marr and T. Poggio. Cooperative computation of stereo disparity. *Science*, 194:283–287, 1976.
13. Gérard Medioni and Ram Nevatia. Segment-based stereo matching. *Computer Vision, Graphics, and Image Processing*, 31:2–18, 1985.
14. S. Pollard, J. Porrill, J. Mayhew, and J. Frisby. Disparity gradient, lipschitz continuity, and computing binocular correspondance. In O.D. Faugeras and G. Giralt, editors, *Robotics Research: The Third International Symposium*, pages 19–26. MIT Press, 1986.
15. S.B. Pollard, J.E.W. Mayhew, and J.P. Frisby. PMF : a stereo correspondence algorithm using a disparity gradient constraint. *Perception*, 14:449–470, 1985.
16. K Prazdny. On the disparity gradient limit for binocular fusion. *Perception and Psychophysics*, 37(1):81–83, 1985.
17. F. Preparata and M. Shamos. *Computational Geometry.* Springer-Verlag, New-York, 1985.
18. L. Robert and R. Deriche. Dense depth map reconstruction: A minimization and regularization approach which preserves discontinuities. In Bernard Buxton, editor, *Proceedings of the 4th European Conference on Computer Vision*, Cambridge, UK, April 1996.
19. S. Russel and P. Norvig. *Artificial Intelligence: A Modern Approach.* Prentice Hall, New Jersey, 1995.
20. D. Scharstein and R. Szeliski. Stereo matching with nonlinear diffusion. *The International Journal of Computer Vision*, 28(2):155–174, 1998.
21. D.S. Sivia. *Data Analysis: a Bayesian tutorial.* Oxford University Press, 1996.
22. D. Weld. An introduction to least commitment planning. *AI Magazine*, 15(4):27–61, 1994.
23. Z. Zhang, R. Deriche, O. Faugeras, and Q.-T. Luong. A robust technique for matching two uncalibrated images through the recovery of the unknown epipolar geometry. *Artificial Intelligence Journal*, 78:87–119, October 1995.

Discussion

1. **Bill Triggs, INRIA Rhône-Alpes**: At each step your matching scheme permanently commits to matches that may be globally suboptimal. How much do you think is lost by this, compared to a scheme with back-tracking, or with optimal look-ahead like dynamic programming?
   Zhengyou Zhang: We do lose if we make a wrong decision. This could happen in areas where there is a discontinuity in depth because my current implementation is mainly based on continuity (the disparity gradient limit principle). Otherwise it will not happen, because based on the least-commitment strategy, I do not make any decision if there is any ambiguity. In terms of number of iterations, my technique may need more than a back-tracking technique; in terms of computation time, I do not know, in fact I have never compared this method to backtracking. The problem with back-tracking is that you have to keep in memory all the previous decisions, and that is not very efficient in terms of implementation. My technique is very simple, and can be easily parallelized.
2. **Andrew Zisserman, University of Oxford**: It is interesting that you use the disparity gradient limit constraint. I would like to know how invariant it is to large camera rotations. Because when it was originally introduced, from the psychophysics literature, they were not considering very severe motions of the camera as it was just for stereo applications.
   Zhengyou Zhang: That is a very good question. In this work I use σ_i to define the uncertainty of the predicted disparity, and it is difficult to know its optimal value, because the disparity gradient limit principle studied in psychology is a beautiful tool for parallel images separated by a fixed distance, say, 10 cm. Here I consider a general configuration so it is an important problem. If σ_i is set to quite a large value many ambiguous solutions can be found and decisions have to be delayed, which implies a slow convergence. If σ_i is set too small, we will not be able to account for enough depth variation and selected matches could be wrong. In the PMF algorithm, K is set to a value between 0.5 and 1, which is equivalent to a value between $1/3$ and $1/2$ for our σ_i. In our implementation, σ_i is a curve, and varies from 0.3 to 1 depending on the distance to a given point match. This is to consider the fact that the disparity gradient constraint is a local one, and that it should become less restrictive when the point being considered is away from a matched point. From the many images I have tried it on, it works quite well. Note that the disparity for any given match is also assumed to be uncertain. Because when a match is selected the precision is limited, an uncertainty of 0.5 pixels is taken into account in the disparity prediction.

Panel Session on Computations and Algorithms

Bill Triggs[1], David Nistér[2], Kenichi Kanatani[3], Jean Ponce[4], and Zhengyou Zhang[5]

[1] MOVI (Modelling for Vision), INRIA Rhône-Alpes, France.
Bill.Triggs@inrialpes.fr
[2] Visual Technology, Ericsson Research, SE-164 80 Stockholm, Sweden.
David.Nister@era.ericsson.se
[3] Dept. of Computer Science, Gunma University, Kyriu, Gunma 376-8515, Japan.
kanatani@cs.gunma-u.ac.jp
[4] Dept. of Computer Science and Beckman Institute, University of Illinois at Urbana-Champaign, USA.
ponce@cs.uiuc.edu
[5] Microsoft Research, Redmond, WA 98052, USA.
zhang@microsoft.com

1 Introduction

The topic of this first panel session was algorithms and computations. Bill Triggs chaired the discussion and David Nister, Kenichi Kanatani, Jean Ponce and Zhengyou Zhang also participated. Each panelist discussed the issues that he felt were going to be important in the future. The panel session was followed by some questions and discussions which are also reported here.

2 Bill Triggs

We asked each panelist to give his views on the following question: What are the most important open areas for research in multi-image algorithms over the next five years?

My own view is as follows. Consider a few typical applications of visual modeling: modeling of buildings or sites, from the interior of a room to city-scale modeling; image-based rendering; and the modeling of human motion and appearance. These are some of the problems for which we would like to be able to build models of some sort from images. These problems have a number of common properties, and I want to emphasize these because I think that the commonality is suggestive of where we are heading.

Firstly, all of these problems are *large, sparse and highly structured*. They have many parameters, but each couples to only a few of the others in an orderly way that reflects the physical structure of the problem. Examples of order are: geometric and temporal locality; causal chains from light source to reflecting surface to camera; visibility constraints that leave only a small proportion of the model visible to any one camera; articulated models of human motion; and Markov state models of scene dynamics. Often there are multiple overlapping levels of such structure. Also, all of these models are *predictive* in the sense that, with some uncertainty, they predict the images or observations from their parameters. To estimate the model, we need to work backwards, inducing parameter values from sets of observations.

Secondly, the models that we need to reconstruct are both *multimodal* and *domain specific*. Pure structure from motion is almost never enough. For graphics applications we need to add appearance and photometric models. For architectural modeling we want higher level geometric primitives (boxes, cylinders), and usually also more semantic information (this room is carpeted with a plaster wall, here is a power point, a door, a light fitting, the desk is a mess so I didn't model it). For human modeling we need to make some sense of a subtly-articulated compound of non-rigid muscle, skin, hair and clothing, to whose minor details we humans are quite extraordinarily sensitive. For scene or motion understanding, we need to add discrete-valued logical state variables, *e.g.* describing action type and phase, scene interpretation in a probabilistic network framework. Often several different sensing modalities are used, so our models need to support more than just conventional camera sensors.

Finally, in all of these problems there is a rich body of *prior knowledge* that must somehow be incorporated into the system to get reasonable performance. Increasingly, this implies some sort of "learning", or more prosaically, prior estimation of background parameters. Recently very interesting results have been obtained by patching or "mosaicing" together learned local appearance models, *e.g.* in face modeling work from A.T.&T., Berkeley and Manchester and motion work from MERL. I think that we will see a lot more of such composites of local models over the next few years. But with or without them, representing, learning and using prior information is still a major problem.

So these are the main areas where I think that general multi-image modeling research could be profitably focused over the next few years: representing, initializing and optimizing large, complex, highly structured models of mixed character (multimodal, both discrete and continuous parameters); extending our expertise on SFM to richer types of models; and learning, representing and using complex prior domain information.

The models that I am thinking of often have a strong semantic component, and in some sense this is a return to the bad old days of AI "scene understanding". But I think that we will see a great deal of progress in many of these applications over the next decade. For one thing, with appearance based approaches, a much improved understanding of structured probabilistic models (HMM's, Bayesian networks), and far more sophisticated structured learning methods, our nonlinear modeling tools are almost immeasurably more flexible and powerful than they were only a decade ago. We are only beginning to tap the potential of this. Moreover, we are in the middle of the wired society boom: the necessary computing power and storage are now there for the asking, and any progress that we make feeds very rapidly into practical applications. In all, I think that computational vision is in for a vibrant, exciting period over the next few years.

3 David Nistér

I have just read Shimon Ullman's book [1] on recognition and I find it very interesting to combine bottom-up with top-down. To obtain a model bottom up from real data and then use the model you have and render it down the pipeline to meet with new bottom-up estimations. But what I want to do is promote the system approach. Maybe you have noticed by now that I am a system person and naturally I will promote this viewpoint.

My experience from building a whole system is the following. There were a number of stages in the development when the results of the components were very discouraging. I was close at these times to concentrating on that component, trying to make it better. But since I was so determined to build the whole system I moved on anyway and what I learned from this is that we can actually make quite decent systems out of components which are not perfect. There will always be outliers. It is just that they have to be taken into account all the time and everywhere in a system. Always expect that your data is bad. I do not want to pin down one thing that would be the research topic for the next five years since I think that that is as difficult as predicting the weather. Instead, I want to propose that some of the things that we have been missing and calling for during the last few years, some of the things that we do not even know that we need, are missing as a symptom of the fact that we are not looking at the full stretch of the problem that we want to solve. A system view might shift the emphasis of the research to new topics as some things do not make sense until you try to build a system. I will give some examples of this:

1. **Synergy effects.** For example, there is a synergy effect between geometry and matching that actually makes matching work. It is not seen when matching is attempted separately from the geometry estimation. Synergy effects like that will only be found when the problem is attacked as a whole.
2. **Accumulation of data.** Accumulation of data is one of the reasons why components with sub-perfect output can still make a good system. For example, if you have two hours of video from the same camera, the information about the calibration is in there, but we can not handle and integrate the sometimes contradictory data from this huge thing. Instead we are often only working on a few images at a time, worrying about degeneracy, which is of course important for a thorough theoretical understanding, but which does not paint a complete picture.
3. **Uncertainty estimates.** Richard Hartley was speaking about uncertainty estimation at ICCV last year and I think he is completely right. This is very important and in all forms of science, something is said about the confidence in a result. Again, I think that the reason why this subject has been lagging in computer vision is that confidence estimates only become necessary when there are other components in a system or a user that demand a confidence estimate. Perturbation analysis is not the whole picture here either, we want to know the cases where the results are a disaster. I would rather have a system that works 50% of the time and tells me that the result is useless

and preferably why, the other 50%, rather than having a system that is right 95% of the times, but presents a disastrous result to the user the other 5%.

4 Jean Ponce

I think that structure from motion has been on the right track. I do not think many fundamental problems remain to be solved. I still think that modeling shape, reflectance and illumination all together is a very hard problem. Illumination is a global process that is extremely hard to analyze and understand. I think that is why people have been focusing on geometry since that is a much more local process. So I think people have to do that. I am kind of naive so I imagine that we still do not deal very well with multiple moving rigid objects, but I may be wrong. I do not think it is very fundamental either. I think on the other hand that one of the interesting points in Paul Debevec's talk was the sense that we are now in the process of moving to applications, but we do not understand the applications very well. We do not understand the market—whatever the market is. People are starting companies, are building products but I am not sure they know who their customers are. I think they should understand what all this stuff is for. We build all these models: what are they for? For the movie industry it is not exactly clear that you want a black box that runs automatically and that will build your model. It is not clear to me at all. I think that at some point we need to acknowledge how—and again this goes back to Paul's talk—to get the person in the loop because I think that a lot of applications want something that always work, rather than something that works in 50% or 60% of the cases. I think it is then quite complicated to understand the user interface process.

5 Kenichi Kanatani

There is one point on which I do not agree with Bill Triggs. He says that we have to integrate prior knowledge in a complicated way. In the 1980s there were lots of discussions about the future of computer vision: people were talking about integrating knowledge in complicated ways. However, recent progress in structure from motion is perhaps because we have avoided involving complicated knowledge. I think that will also be true for some time to come.

Now, I want to make a different point. Nowadays, structure from motion means multiple image reconstruction thanks to today's computing power, and effective 3D reconstruction techniques have successfully been developed based on geometric constraints, such as rigidity, planarity, and various camera models, that govern the images. This means that for 3D reconstruction we need to know geometric constraints to exploit, but we are not always sure if the geometric constraint that we impose is correct, which among possible constraints really exists, or if the constraint happens to be degenerate. As is well known, degeneracy frequently occurs even when the camera motion is very natural, and we cannot retrieve the 3D shape in such critical configurations. In this context, model selection emerges as a new challenge.

I have been studying this problem for some time, and I have realized that we cannot always use stochastic model selection criteria, by which I mean those found in textbooks on statistics. Textbooks on statistics are written by statisticians, who deal with traditional statistics: they talk about Akaike's AIC, Rissanen's MDL, Schwarz' BIC, and other criteria. My conclusion is that 3D reconstruction is not a statistical problem. It appears to be a statistical problem, but it is a geometric problem. We must make a distinction between statistical inference and geometric inference.

In statistical inference, the accuracy of estimation increases as the number of observations increases. So, if we are asked how we can maximize the accuracy of estimation for a limited number of observations. The answer is to choose the one for which accuracy increases most rapidly as the number of observations increases. In geometric inference, on the other hand, we are dealing with errors and noise, and we are interested in maximizing the accuracy of estimation for limited resolution. So, we choose the one for which accuracy increases most rapidly as the noise level decreases. Thus, we need a new model selection theory different from the traditional one, which only very few people seem to have realized. This may be one of our main challenges.

6 Zhengyou Zhang

I will talk more on the algorithmic aspects instead of research topics. I think vision is currently at a stage where it can be useful in many applications. There are several factors which contribute to this achievement. One thing is that when we develop algorithms we should take into account the uncertainty of the data. This is very important. Twenty years ago people usually looked for linear algorithms which discarded the noise property of data, and they did not give good results. About fifteen years ago, people realized the importance of taking account of data uncertainty, and both analytical and nonlinear algorithms have been designed which give much superior results. So to keep vision successful we need to take into account the data uncertainty. Bundle adjustment is a good example of this: it says basically that image points are detected with similar uncertainty and we should minimize an error function defined in the image space. Gradient-weighted least-squares is another example.

The second success factor consists in developing robust techniques, because there are always outliers present in the data. RANSAC, M-estimators, LMedS are becoming standard tools. We should certainly continue to use them. There are also a lot of studies on the stability of the algorithms, i.e. degenerate configurations. At the moment this is carried out on the noise-free situation. So in the future we need to study the stability in the case where the data is noisy. Another factor that I want to mention is that the vision algorithms are successful when we incorporate, as much as possible, prior knowledge, e.g. special camera models, domain knowledge or parallelism. I see several topics that will become increasingly important in the future: how to systematically asses the usefulness of the obtained results and how to detect systematically when an algorithm fails and suggest what to do alternatively. Because vision is often used as a component or module of a larger system, we need to know when it fails and how to do things differently.

The third point is how to systematically improve the algorithm when it fails. When the algorithm fails it should learn from that failure in an automatic, intelligent, way so that a better algorithm is obtained. The last point I want to mention is that we should try to automatically incorporate prior knowledge instead of hard coding it. Visual learning is a very useful and important area to explore.

Discussion

1. **Daniel Cremers, University of Mannheim**: I have a question about the different methods to reconstruct 3D structure, mainly in how you can compare the different results that they give. I have the impression that you have two steps. First you reconstruct the scene and then you map texture on it. Then the quality always seems to be mostly determined by visual inspection. I have the impression that these steps, especially the texture mapping, somewhat occludes the results of comparing the methods. How do you go about comparing them?
Rick Szeliski, Microsoft: I think first of all we have to specify what the problem domain is. If the application would be robotics, the goal is to not run into things or break stuff. Let us assume that we are working on the general category of pretty things we do with image, in which I would include visual effects you have in movies and image-based rendering. Then, say we constructed a model, the final output you want is for it to look acceptable. The measure of success is that you have produced something that would be acceptable to an audience watching a movie. So it has to be basically visually perfect. There is a systematic way of doing that. If you take a large collection of images and you hold some of the images out of the reconstruction (this is what people in machine learning have been doing for decades with great success) you can basically evaluate the quality of the reconstruction by testing against the images you have held out. This test can be used both for interpolating or extrapolating. The one open problem I think we do not know how to solve yet is how to accurately model visual quality and perception. I do not think we have good models of that. That is one potential answer to your question.
Jean Ponce: If I may comment on your question. I agree with you. For those of us who have been around for a fairly long time: stereo used to stink. The results were awful. Then, starting in the mid-eighties, people started to texture map the results and suddenly stereo looked beautiful. The results are just as bad as they used to be, but when you paint a picture on the surface it looks good. There is a problem with that.

Rick Szeliski: Why is it that it is a bad result if it looks beautiful? At least if you do not try avoiding running into something.

Jean Ponce: It depends on the application, of course. At the time it was not clear at all what the stereo applications were. It was done mostly for the sake of it.

Daniel Cremers: What I want to criticize is the system aspect. If you put a lot of modules together and then look at the final result, you can not tell how good each step is. I assume that for example various people use the same texture mapping routine, so they should compare the results before doing that.

David Nister: What I wanted to say was not that everybody here should take the system approach. We would get nowhere then. I am just saying that it is good if everybody has a wider perspective and knows where their research comes into the big picture. And also, we all know that vision is inference with uncertainty. There will always be bad results for some kinds of input and you might end up banging your head at some problem where you can not really get better results (not that I think we are there yet). And then I would also like to add to what Rick says: you need to take out views to verify with those. I think that provided that you have restricted your model reasonably, in many cases it is not even necessary to take out views, you can just try to reproject your views and that is difficult as it is.

Hans-Helmut Nagel: I just want to comment on that. I take the opposite point of view. If you evaluate a component in a system environment, you can modify the component and the system reaction is a much more appropriate assessment than if you make up an individual test environment for that component. So I would rather go for the system and modify the component, check its reaction than testing each individual component in its own testing environment.

2. **Henrik Aanaes, Technical University of Denmark**: Maybe I have misunderstood something, since I am a bit of a novice in this. Prof. Kanatani seems to see geometry and statistics as two distinct things. My intuition would be to see geometry in a statistical setting, because then you would also be able to have a much better evaluation and a much better understanding of the stability of your solution. More than you would be able to get from your perturbation analyses. And then you would be able to be in David Nister's ball park. Maybe that would be where you would be able to see large uncertainties on your solutions and thus be able to infer if you actually had a stable solution.

   Kenichi Kanatani: Yes, that is right. What I wanted to point out is that you have to have a statistical sense and analyze geometric problems with statistical principles, but so far people are too slow to understand this, merely interested in picking out methods from textbooks. My message is that you should rather throw away textbooks and think about the problem on your own.

3. **Rick Szeliski**: Unless somebody wants to continue along these lines I want to introduce a new point. It is actually a restatement of one of the points

that Bill Triggs made. I love working in this field, I think we have great results, but I am struck from time to time how still and dead our worlds are. Everything we reconstruct is static. And yet a lot of the action out there in terms of graphics and that is character animation and things like dynamic visual effects. There are two ways of going after dynamic models. You can go build a large video rig, like those which Kanade and other people have made. I think that is going to be a fruitful area of research since you don't want to get just a collection of independent static models. The other one that is more challenging—and I throw it as an open gauntlet since I am not sure I can even solve it—is to take a moving camera in a moving world and see how much of it you can reconstruct. Take a video camera walking down the streets of Dublin and come back to me in three years and show me how much of that you have reconstructed. That is certainly an open challenge to our community.

4. **Zhengyou Zhang**: I would like to say a few words about an open challenge, related to the first question, about the evaluation of the algorithms. This needs a common database of software. Everybody should publish his/her software (at least in executable form) in order to allow others to try it on more data. If you are the only user you can just tune it to a small set of data and get a very good result, but this does not make sense. The algorithm should be verified on a variety of data. We would thus also need to have some image databases to compare our algorithms on.

5. **Jean Ponce**: A last challenge is maybe that apparently there are really good range finders that are really cheap and that give you 512 × 512 pictures in real-time with millimeter type resolution. And so, what is going to happen to this community when those come around? I must say honestly, of course there is already existing footage that you want to analyze, but if you can really buy for only $50 an add-on for your video camera that does the job for you. So I don't know but I think it is interesting to see what people think about these things.

Paul Debevec: I think that in response to that there are still going to be a lot of issues involved once we have the "Zcam" that 3DV Systems is working on in Israel. I will be extremely excited when that is available, but there are still a lot more issues that are to be solved which have to do with the reflectance, photometry, registering all of the views, dealing with noise, integrating it with things that are beyond the range the camera can recover (i.e. the deeper environment). Hopefully this is the community right here that can answer those challenges.

References

1. Shimon Ullman. *High-Level Vision, Object Recognition and Visual Cognition*. MIT Press, ISBN 0-262-21013-4, 1996.

Rendering with Non-uniform Approximate Concentric Mosaics

Jinxiang Chai[1], Sing Bing Kang[2], and Heung-Yeung Shum[3]

[1] The Robotics Institute, Carnegie Mellon University,
Pittsburgh, PA 15213, USA
[2] Vision Technology Group, Microsoft Research,
Redmond, WA 98052, USA
[3] Microsoft Research,
Beijing, China

Abstract. In this paper, we explore the more practical aspects of building and rendering concentric mosaics. First, we use images captured with only *approximately* circular camera trajectories. The image sequence capture can be achieved by holding a camcorder in position and rotating the body all around. In addition, we investigate the use of variable input sampling and fidelity of scene geometry based on the level of interest (and hence quality of view synthesized) on the objects in the scene. We achieve the tolerance for minor perturbations about the exact circular camera path and variable input sampling by using and analyzing a variant of the Hough space of all captured rays. Examples using real scenes are shown to validate our approach.

1 Introduction

Image-based rendering (IBR) has become a popular approach for modeling and rendering a virtual environment. While the conventional means of rendering uses a 3D model (with possibly a complicated photometric model), image-based rendering directly interpolates novel views from captured images. If the input images are captured sparsely in the space, establishing correspondences may still be necessary. However, if the input images are densely captured, direct view interpolation will suffice.

In theory, one needs only to capture a complete plenoptic function [1,7] in order to synthesize a novel image from any viewpoint and at any viewing direction. However, a complete plenoptic function is at least 5D, which includes 3D spatial location and 2D ray directions at any point. If free space is assumed, the plenoptic function can be reduced to 4D, as shown in the lumigraph [2] and light field rendering [6]. However, for modeling a virtual environment, the size of the database for the light field is usually massive because it has to sample four dimensions.

Recently, concentric mosaics [11] has been proposed to sample a virtual environment where the viewpoints are constrained on a planar surface. It has been shown in [11] that a novel view can be generated from a sequence of images captured from a camera rotated off-center along a circular path. A *linear pushbroom camera model* is assumed [3] (as is with our work). In other words, the camera model used comprises a stack of parallel perspective views perpendicular to the y-axis, with each perspective view representing a horizontal scanline. While vertical distortion exists as a result of using this camera model, the synthesized images show good rendering quality with the help of constant depth correction and bilinear interpolation.

However, there are at least two disadvantages associated with the current concentric mosaics work. First, it requires a capturing rig that is bulky. It is much more practical if a user can hold a camcorder in a position and rotate his body around to capture the necessary images. Second, it is desirable to capture the environment with variable sampling rates and fidelities. For example, it is intuitive that more samples should be taken at regions that are deemed more interesting. It also makes more sense to make more samples at areas that is highly textured and where depth variation is significant.

This paper addresses the above two practical issues in concentric mosaic building and rendering, namely using hand-held camera to acquire images and variable input sampling. The input sequences of images are captured using a hand-held camera, and recovery of the camera pose is accomplished using a structure from motion algorithm. However, we do not explicitly build a 3D model from the input images (e.g., generate 3D panoramic models from stereo [4]). To handle the variable sampling resolution, we propose a new representation we call called *signed Hough space* that enables uniform sampling and efficient computation in the ray space.

1.1 Previous Work

There has been significant work done on image-based rendering using large quantities of input images. The pioneering work on the lumigraph [2] and light-field rendering work [6] have spawned a number of related work. Two of the more notable ones are the concentric mosaic [11] and the stereo panorama [8]. There are also others who use the approach of generating 3D panoramic models [4], or computing panoramic depth as a means for rendering [7,12].

1.2 Outline of Paper

The remainder of this paper is organized as follows. We describe our new representation called *signed Hough space* in Section 2. In Section 3, we give a summary of the least-squares method to extract camera pose from a sequence of tracked images. Once camera poses are known, the input data is mapped to the new representation space. Issues with rendering with approximate concentric mosaics using the new representation is discussed in Section 4. Experimental results using synthetic and real images are shown in Section 5. We conclude this paper in Section 6.

2 Signed Hough Space

Our image-based approach is based on reusing captured rays from input images to reconstruct an image at a novel viewpoint. An important problem in image-based rendering is the representation, namely, how to represent the rays that are captured. For example, the lumigraph is a particular way of sampling the ray space using a 4D two-plane parameterization. Concentric mosaics sample the space using three parameters, i.e., the rotation angle, radius and vertical field of view. In this section, we present a new approach to represent non-uniform concentric mosaics from a large collection of images taken along an approximate circle. The major issue in choosing a representation for non-uniform plenoptic sampling is how to parameterize the space of oriented lines. We consider a good choice of parameterization of oriented rays to have the following characteristics:

- *Efficient calculation.* The computation of the position of oriented ray from its parameter space, and vice versa, should be fast.
- *Uniform sampling.* The sampling within the spatial and directional spaces should be uniform. This is to avoid potential problems in rendering.
- *All inclusive.* All possible oriented rays in the space should be represented, with no exceptions.

Note 1.
Duality. Reciprocal behavior should exist between the destination (within a panorama in view space), and source (a geometric point with its radiance in Cartesian space). In other words, analysis would proceed exactly the same if the destination and source are switched. It is obvious that light field representation using the two-plane parameterization cannot satisfy the third item. Rays that are parallel or do not intersect the slabs are not represented. In our case, rays at all orientations and positions can be included in our representation.

Note 2. For simplicity, we first describe the representation of oriented rays in 2D Cartesian space, and then we will extend it to 3D space for the representation of approximate concentric mosaics. One of the ways that we can visualize the population of rays available is to construct the usual Hough space which uses the normal (r, θ) parameterization. However, rays are directional, and the conventional Hough space is unable to distinguish rays that have the same equation by are of opposite directions. We solve this by using the right-hand rule: A ray that is directed in an anti-clockwise fashion about the coordinate center is labeled positive, otherwise it is labeled negative. "Positive" rays have positive r values, i.e., (r, θ), while "negative" rays have negative r values, i.e., $(-r, \pi + \theta)$. Figure 1 shows four different rays in 2D space and their corresponding points in the signed Hough space.

An attractive feature of this representation is the duality between points and sinusoids in both Cartesian and signed Hough space. Figure 2 shows examples of common projections are represented in signed Hough space. For example, panoramic visibility at a point in Cartesian space (Figure 2(a)) is represented as a sampled sinusoidal curve in the parameter space. A concentric mosaic (Figure 2(b)) is mapped to a horizontal line in the signed Hough space, while parallel projections (Figure 2(c)) are mapped to a vertical line in the signed Hough space.

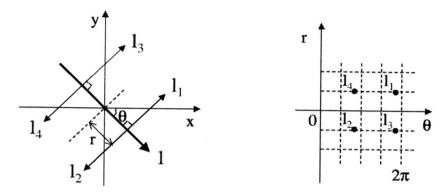

Fig. 1. Definition of the ray space we captured to reconstruct the 3D geometry. Each oriented ray in Cartesian space (at left) is represented by a sampled point in the signed Hough space.

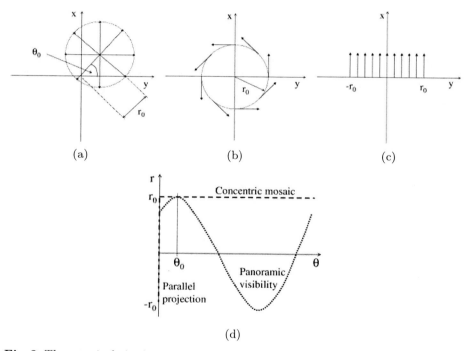

Fig. 2. Three typical viewing setups and their associate sampled curve in signed Hough space. (a) Panoramic visibility at a point in 2D Cartesian space, (b) A concentric mosaic, (c) Parallel projection, and (d) Their respective sampled curves in the signed Hough space.

Note 3. Specifically, the bundle of all rays emitted by a 3D geometric point in Cartesian space also takes the shape of a sampled sinusoidal curve featured by its space location (r_0, θ_0). Thus, the captured perspective scene can be easily transformed into the parameter space. Rendering a novel view in the scene is equivalent to extracting a partial sinusoidal curve from the signed Hough space. Interestingly, computing the depth of scene can also be defined as a curve fitting problem that is constrained by a specific BRDF model.

3 Rendering Using Handheld Sequential Images as Input

The previous work on concentric mosaic [11] uses images from a camera with a perfectly circular trajectory using a motorized setup. We extend this work to a more practical level by allowing visualization from *approximate* concentric mosaics. The input images can be captured from a hand-held camera that is moved through an approxiately circular trajectory.

3.1 Computing Structure from Motion

Building the approximate concentric mosaic requires accurate camera poses associated with the input images. To do this, we first calibrate the camera to extract intrinsic parameters using the method described in [15]. Subsequently, we automatically track point features in the image sequence using Shi and Tomasi's tracker [10]. Their tracker uses an affine model and a Hessian-based measure of the local texturedness to determine removal and addition of point features at each frame.

Once the point tracks are available, we apply the iterative least-squares minimization technique based on Levenberg-Marquardt on these point tracks [14] to recover camera motion. For completeness, we provide a brief description of this algorithm.

Structure and motion are solved simultaneously to minimize the difference between the 2-D track points and the 3-D object points projected into 2-D. The Levenberg-Marquardt algorithm [9], a standard iterative least-squares solver, is used to minimize the objective function

$$\mathcal{C}(\mathbf{a}) = \sum_i \sum_j c_{ij} |\mathbf{u}_{ij} - \mathbf{f}(\mathbf{a}_{ij})|^2, \qquad (1)$$

where $\mathbf{u}_{ij}$ is the measured point feature location, $\mathbf{f}(\mathbf{a}_{ij})$ is the predicted projected point,

$$\mathbf{a}_{ij} = (\mathbf{p}_i^\mathbf{T}, \mathbf{m}_j^\mathbf{T}, \mathbf{m}_g^\mathbf{T}) \qquad (2)$$

and c_{ij} is a measure of confidence of the position, based on the amount of local texture at the point.

The vector $\mathbf{a}$ contains the 3-D points $\mathbf{p}_i$ for each point i, the local motion parameters $\mathbf{m}_j$ for each frame j, and the global motion and camera intrinsic parameters $\mathbf{m}_g$. The function $\mathbf{f}(\mathbf{a}_{ij})$ is the projective function that maps the point $\mathbf{p}_i$ to the image j, using the camera position and the camera intrinsic parameters.

For each iteration, the Levenberg-Marquardt algorithm finds an approximate Hessian matrix $\mathbf{A}$ and gradient vector $\mathbf{b}$, which is used to solve for an increment $\delta \mathbf{a}$ towards the minimum. The equation solved is

$$(\mathbf{A} + \lambda \mathbf{I})\delta \mathbf{a} = -\mathbf{b}, \tag{3}$$

where λ is a time-varying stabilization factor and $\mathbf{I}$ is the identity matrix.

The elements of the Hessian $\mathbf{A}$ are approximated as the product of partial derivatives with respect to $\mathbf{a}$:

$$\mathbf{A} = \sum_i \sum_j 2c_{ij} \frac{\partial \mathbf{f}^T(\mathbf{a}_{ij})}{\partial \mathbf{a}_{ij}} \frac{\partial \mathbf{f}(\mathbf{a}_{ij})}{\partial \mathbf{a}_{ij}^T}, \tag{4}$$

and the gradient vector $\mathbf{b}$ is

$$\mathbf{b} = \sum_i \sum_j 2c_{ij} \frac{\partial \mathbf{f}^T(\mathbf{a}_{ij})}{\partial \mathbf{a}_{ij}} \mathbf{e}_{ij}, \tag{5}$$

where $\mathbf{e}_{ij} = \mathbf{u}_{ij} - \mathbf{f}(\mathbf{a}_{ij})$ is the position error.

Note 4. For our application of rendering with approximate concentric mosaics, we would also like to constrain the camera motion to a simple planar motion from general rigid motion. The structure from motion algorithms would be more robust with the reduction in the number of parameters.

Once we have obtained the camera poses using the tracker and subsequent structure from motion algorithm, we can then map all the input rays associated with the cameras to the signed Hough space for subsequent rendering.

4 Rendering from the Signed Hough Space

By resampling the input rays into the signed Hough space, we can achieve the tolerance for minor perturbations about the exact camera poses. These camera parameters may not be perfectly recovered from the above structure from motion algorithms. In the new space, we improve rendering quality by designing optimal interpolation filters. We analyze various interpolation filters, including parallel interpolation and constant depth interpolation along r and θ directions. Furthermore, multi-resolution rendering (i.e., zoom in and out of objects/regions of interest) can also be easily implemented in the new representation space.

Given a set of non-uniform concentric mosaics collected from a camera moving non-uniformly along an approximately circular path, we can render any novel view. The rendered views are constrained by the camera trajectory, similar to concentric mosaics where viewpoints of the rendering camera are constrained by the capturing circle.

Rendering a new image at any viewpoint becomes the problem of extracting a sinusoidal curve in the signed Hough volume. However, due to the discretization of the signed Hough volume, interpolation techniques have to be carefully chosen in order to obtain high quality rendering results.

Before we describe the interpolation techniques, let us make a couple of definitions, with the help of Figure 4. All the rays for a given virtual camera

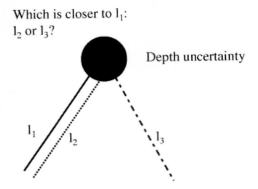

Fig. 3. Ambiguous definition of closest ray.

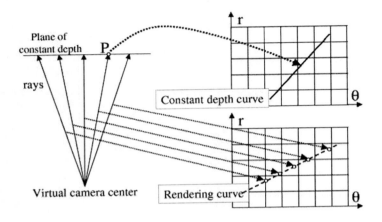

Fig. 4. Rendering and depth correction curves.

map into what we call a *rendering curve*. If the depth correction is specified, any given ray will intersect at a known point, say P. P then maps onto the *depth correction curve* in ray space.

To continue, a good interpolation filter should make use of depth information. However, when no information about the scene geometry is available, the parallel bilinear filter (e.g., [11]) is commonly used to interpolate the rendering rays. It works by assuming all of the scene points are located in infinity, as shown in Figure 5(a). In this particular case, the four closest ray bins I_1, I_2, I_3, and I_4 are used to compute the color of the virtual ray indicated by $\tilde{I}_{m,n}$.

Bilinear interpolation and constant depth assumption can be used to improve the quality of rendered images. With the constant depth assumption, all of the objects seen by the camera are deemed to be located along a simple surface such as a cylinder. As with any assumption on scene depth, the issue is how to choose the closest points to reconstruct the rendered point.

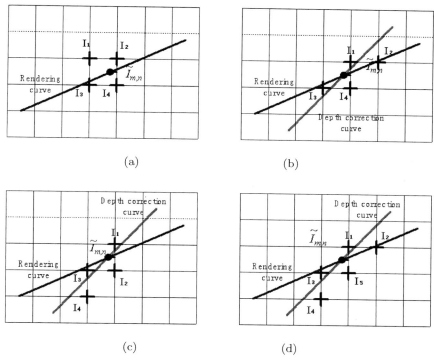

Fig. 5. Different bilinear interpolation filters. (a) Parallel bilinear interpolation, (b) Bilinear interpolation with constant depth correction along angular direction, (c) Bilinear interpolation with constant depth correction along radius direction, and (d) Bilinear interpolation with constant depth correction along both directions. Note that the horizontal axis is that of θ while the vertical axis is that of r.

The definition of "closest" points is ambiguous if no accurate depth information is known. Consider, for example, the question as to which of the rays, l_2 or l_3, is "closer" to ray l_1? The notion of closeness makes sense only if the object distance is known, even approximately. The interpolation techniques shown in Figure 5(b)-(d) uses specified depth corrections to decide which ray bins to use. As an example as to how the ray bins are chosen for interpolation, consider the case of constant depth correction along the angular direction, as shown in Figure 5(b). First, the intersections between the depth correction curve and horizontal rows closest to the virtual ray $\tilde{I}_{m,n}$ are computed. The sampling ray bins are those just on each horizontal side of these intersections. Similar reasoning can be applied to Figure 5(c) and (d).

5 Experiments

Unlike most capture setups for image-based rendering, the image capture process here is very simple. Specifically, a single camera is moved by hand to rotate along an approximate circular path. In our experiments, a total number of 1864 images of a real scene is captured. The image size is 360 × 288. Only 530 frames are used to recover camera poses using our SFM algorithm. Two input images are shown in Figure 7(a)(b) where a number of feature points are tracked for the SFM algorithm. As shown in Figure 6, the rotation and translation parameters are recovered fairly well.

Using the estimated camera motion, we transform the input images into our signed Hough space. The binning process is based on nearest neighborhood. The new parameter space has the resolution of 230 × 310 in radial and angular dimensions. The signed Hough space can also be examined to see if it can be represented with coarser discretization by checking the density of ray occupancy. Downsampling has the benefit of compactness. In addition, we have applied vector quantization compression to our database to further reduce its size; in our example, the reduced size is about 4MB.

Figure 7(c,d) show two rendered images. Note the significant parallax changes around the monitor in the middle and through the window on the right. Four different interpolation techniques have been applied to render the new images, as shown in Figure 8. These techniques are parallel interpolation, depth correction around radial direction, depth correction around angular direction, and depth correction with both radial and angular directions, respectively. Among these techniques, depth correction along radial direction produces the best rendering result, whereas depth correction along angular direction is the worst. Because angular sampling is much denser than radial sampling in the original images, interpolation along radial direction is effective. In fact, the angular direction is over-sampled. Depth correction along both directions produces comparable rendering result as with depth correction along radial direction only. Parallel interpolation has better rendering result than depth correction along angular direction because parallel interpolation is in fact along the radial direction, albeit at the infinite radius.

With the new parameter space, we can also render images in different resolutions. Figure 9 shows the results of zooming in and zooming out. Notice the appropriate changes in apparent size of the bunny. In general, there are two approaches to obtain the zoom-in effect. First, we can sample the areas of interest more densely than others. But multi-resolution representations should be applied for efficiently storing the data. Second, depth information can be used to improve the resolution. Higher resolution of output images can be achieved with more accurate depth information. The depth information can be obtained by either vision reconstruction techniques or human interaction. For example, Figure 9(b) is obtained with a different depth specified by the user than the depth used in Figure 9(a).

6 Discussion

Database acquisition for light-field-based IBR is usually a very laborious process and often require specialized (and thus expensive) equipment. Until drastic simplications are made to the acquisition process, IBR will remain beyond the reach of ordinary consumers. With our technique, however, such specialized equipment is not necessary. We have shown that we can provide high-quality visualization

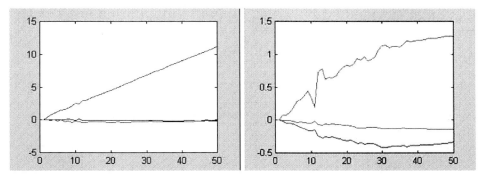

Fig. 6. Camera poses estimated using structure from motion algorithms. Left: Graph depicting the variation in rotation (in degrees) about the y, x, and z axes (curves from top to bottom). Right: Graph depicting the variation in translation along the x, y, and z axes (curves from top to bottom).

from a database created from images taken using just a hand-held camera that is manually moved along an approximately circular path.

We have also used the notion of variable sampling in our work. In areas where objects are less interesting to us, we can afford sparser input sampling and without (or with less accurate) depth information. This may not be very evident in our results, because our overall sampling is actually rather dense, even in the least densely sampled areas.

While the camera motion parameters are required to build the database for the concentric mosaic, absolute accuracy of these parameters are, in practice, not necessary. This is evidenced by our results. There are enhancements to our current SFM algorithm that we can make. Our SFM algorithm is currently too general. If we know that the motion is planar (or assumed planar), we can impose additional constraints in our algorithm, so that fewer parameters need to be computed. (In the handheld camera case, this may or may not be applicable.) Parameter recovery will be faster as well, especially when we are dealing with a large number of images and tracks.

7 Conclusions and Future Work

In this paper, we have proposed a practical method for capturing and rendering approximate and non-uniform concentric mosaics. The method does not require a specialized rig for image capture; manually moving a hand-held camera along an approximately circular path is sufficient. In addition, we introduced the *signed Hough space* to represent the captured rays. The extension to the conventional Hough space is necessary in order to encode rays with direction. For full 3D space of rays (i.e., using a normal perspective camera model instead of a pushbroom camera model), we can use an alternative representation based on *oriented projective geometry* [13]. This representation has been used to recover shape from silhouettes [5].

Fig. 7. Rendering with non-uniform concentric mosaics. (a,b) Two frames in the input image sequence, and (c,d) Two rendered images with significant parallax change.

Judicious use of variable input sampling can be effective in making more optimal use of the available limited manual and rendering resources. This basically trades off fidelity of output with the level of interest. We intend to investigate this aspect more thoroughly.

Finally, we have describe different interpolation regimes and show the results of applying them. The bilinear interpolation with depth correction seems to work the best.

References

1. E. H. Adelson and J. R. Bergen. *Computation Models of Visual Processing*, chapter The plenoptic function and the elements of early vision. MIT Press, Cambridge, MA, 1991.

Fig. 8. Results of using different bilinear interpolation filters. (a) Parallel bilinear interpolation, (b) Bilinear interpolation with constant depth correction along angular direction, (c) Bilinear interpolation with constant depth correction along radius direction, and (d) Bilinear interpolation with constant depth correction along both directions.

Fig. 9. Results of zooming in and out. (a) No zoom, (b) Zooming in with a factor of 0.75, and (c) Zooming out with a factor of 1.25. Note the size of change of the bunny.

2. S. J. Gortler, R. Grzeszczuk, R. Szeliski, and M. F. Cohen. The lumigraph. *Computer Graphics (SIGGRAPH'96)*, pages 43–54, August 1996.
3. R. Gupta and I. H. Richard. Linear pushbroom cameras. *IEEE Transactions on Pattern Analysis and Machine Intelligence*, 19(9):963–975, September 1997.
4. S. B. Kang and R. Szeliski. 3-d scene data recovery using omnidirectional multibaseline stereo. *International Journal of Computer Vision*, 25(2):167–183, November 1997.
5. K. N. Kutulakos. Shape from the light field boundary. In *IEEE Computer Society Conference on Computer Vision and Pattern Recognition*, pages 53–59, Puerto Rico, June 1997.
6. M. Levoy and P. Hanrahan. Light field rendering. *Computer Graphics (SIGGRAPH'96)*, pages 31–42, August 1996.
7. L. McMillan and G. Bishop. Plenoptic modeling: An image-based rendering system. *Computer Graphics (SIGGRAPH'95)*, pages 39–46, August 1995.
8. S. Peleg and B. Ben-Ezra. Stereo panorama with a single camera. In *IEEE Computer Society Conference on Computer Vision and Pattern Recognition*, pages 395–401, Fort Collins, CO, June 1999.
9. W. H. Press, B. P. Flannery, S. A. Teukolsky, and W. T. Vetterling. *Numerical Recipes in C: The Art of Scientific Computing*. Cambridge University Press, Cambridge, England, second edition, 1992.
10. J. Shi and C. Tomasi. Good features to track. In *IEEE Computer Society Conference on Computer Vision and Pattern Recognition*, pages 593–600, Seattle, WA, June 1994.
11. H.-Y. Shum and L.-W. He. Rendering with concentric mosaics. *Computer Graphics (SIGGRAPH'99)*, pages 299–306, August 1999.
12. H.-Y. Shum and R. Szeliski. Stereo reconstruction from multiperspective panoramas. In *International Conference on Computer Vision*, pages 14–21, Kerkyra, Greece, September 1999.
13. J. Stolfi. Oriented projective geometry. In *Annual Symposium on Computational Geometry*, pages 76–85, Waterloo, Canada, June 1987.
14. R. Szeliski and S. B. Kang. Recovering 3D shape and motion from image streams using nonlinear least squares. *Journal of Visual Communication and Image Representation*, 5(1):10–28, March 1994.
15. Z. Zhang. Flexible camera calibration by viewing a plane from unknown orientations. In *International Conference on Computer Vision*, pages 666–673, Kerkyra, Greece, September 1999.

Discussion

1. **Kyros Kutulakos, University of Rochester**: I have a couple of comments regarding related work. There was a paper by Wright et al [3]. They use something very similar to the signed Hough space which seems to be quite related. Also I should point out that this idea of the signed Hough space is actually closely related to oriented projected representations of the space of lines. Mainly, any line on the plane can be projected onto the oriented projective sphere. The sinusoids, that you describe, map to great circles on that sphere. This has two advantages over the representation that you describe. First of all, it is not sinusoids but great circles, which makes for a more structured distribution of points or pixel values over the sphere. The

other advantage is that you can use that representation even if you don't have exact calibration. The Hough space representation requires that you know the angles. You can create these lightfields purely projectively as long as you know the projective calibration of your camera. This is something we investigated in CVPR'97, how you can actually represent slices of lightfield on a single epipolar plane by mapping pixels in that lightfield onto the oriented projective sphere.

Sing Bing Kang: Those are good points. I am not aware of the first work that you have just mentioned, and I appreciate your pointing that out. To address the second part of your question, each row basically has a dimensionality of two. I think that Hough space in 2D is more compact than a 2D (spherical) manifold in 3D.

Kyros Kutulakos: That sphere can be mapped stereo-graphically onto a plane as in Geyer and Daniilidis [1]. That allows you to map points to lines and lines to points.

Sing Bing Kang: Yes, but then the stereographic transformation that maps a sphere to a flat 2D surface is non-uniform.

Kyros Kutulakos: That is true. I'm just saying that when you describe things on the sphere you can use a spherical quadtree or something similar to better get a handle on the structure of the space. When you do things on a plane you have indeed these warpings.

Sing Bing Kang: We want the interpolation to be accomplished in a uniform manner in whatever parametric space we choose. Using the sphere and stereographic projection would lead to non-uniform grids, and so, to us, it may not be that effective.

2. **Richard Szeliski, Microsoft**: You described the interpolation, but you didn't say how the original images or rays are put into your Hough data structure. Is there resampling involved?

    Sing Bing Kang: Yes, and the resampling is based on just the closest point. In other words, we use bins to store the colour of the rays, and each sampling ray is mapped to the closest bin. Rays that happen to map to the same bin have the average of their colours stored instead.

3. **Bill Triggs, INRIA Rhône-Alpes**: Just a comment. The camera model that you're assuming, with affine layers vertically but perspective projection horizontally, is called a linear pushbroom camera. I'm not sure whether it will help you, but you can read about it in Gupta and Hartley [2].

4. **Paul Debevec, University of Southern California**: In the demo you said you were zooming the camera in and out. Were you actually physically moving the camera in and out or were you just changing the focal length? When I hear of zooming, I think of just changing the focal length.

    Sing Bing Kang: Yes, the focal length is merely changed under zooming.

    Paul Debevec: So there is no parallax going on there. So you're just making the image that we are seeing bigger and smaller.

    Sing Bing Kang: There is actually a mode in the demo where you can translate forwards and backwards, but the problem is that you cannot see much of the resulting parallax. That is why I did not show it. I have instead demonstrated the effect of translating sideways.

    Paul Debevec: I just couldn't quite tell if there was parallax or just aliasing that made it look like parallax.

Sing Bing Kang: There is a pure zooming mode which does not provide any parallax; as I have mentioned before, there is also another mode which allows you to translate forwards and backwards. In the latter mode, you should get parallax, but not much. That is why I did not demonstrate this latter mode.

References

1. C. Geyer and K. Daniilidis. A unifying theory for central panoramic systems and practical implications. In *Proc. European Conference on Computer Vision*, pages 445–461, 2000.
2. R. Gupta and R. I. Hartley. Linear pushbroom cameras. *IEEE Transactions on Pattern Analysis and Machine Intelligence*, 19(9):963–975, 1997.
3. M. Wright, A. W. Fitzgibbon, P. J. Giblin, and R. B. Fisher. Convex hulls, occluding contours, aspect graphs and the Hough transform. *Image and Vision Computing*, 14(8):627–634, 1996.

Volumetric Warping for Voxel Coloring on an Infinite Domain

Gregory G. Slabaugh[1], Thomas Malzbender[2], and W. Bruce Culbertson[2]

[1] School of Electrical and Computer Engineering
Georgia Institute of Technology
Atlanta, GA 30332
slabaugh@ece.gatech.edu
[2] Client and Media Systems Lab
Hewlett-Packard Laboratories
Palo Alto, CA 94306
{tom_malzbender, bruce_culbertson}@hp.com

Abstract. Starting with a set of calibrated photographs taken of a scene, voxel coloring algorithms reconstruct three-dimensional surface models on a finite spatial domain. In this paper, we present a method that warps the voxel space, so that the domain of the reconstruction extends to an infinite or semi-infinite volume. Doing so enables the reconstruction of objects far away from the cameras, as well as reconstruction of a background environment. New views synthesized using the warped voxel space have improved photo-realism.

1 Introduction

Voxel coloring algorithms [7] [5] [2] reconstruct three-dimensional surfaces using a set of calibrated photographs taken of a scene. When working with such algorithms, one typically defines a *reconstruction volume*, which is a bounding volume containing the scene that is to be reconstructed. Once defined, the reconstruction volume is divided into voxels, forming the voxel space in which the reconstruction will occur. Voxels that are consistent with the photographs are assigned a color, and inconsistent voxels are removed (carved) from the voxel space [7].

These algorithms have been particularly successful in reconstructing small-scale scenes that are restricted to a finite domain. Applying them to large-scale scenes can become challenging, since one must use a large reconstruction volume to contain the scene. Such a large reconstruction volume can consist of an unwieldy number of voxels that becomes prohibitive to process. In addition, it is unnecessary to model far away objects with high resolution voxels. Ideally, one would like a spatially adaptive voxel size that increases away from the cameras.

Furthermore, voxel coloring algorithms are not well suited to capturing the environment (sky, background objects, etc.) of a scene. Typical reconstructions are photo-realistic in the foreground, which is modeled, but empty in the background, which is unmodeled. As a result, synthesized new views can have large "unknown" regions, as shown in black in Figure 1. For some scenes, such as an outdoor scene, we might like to reconstruct the background as well, yielding a more photo-realistic reconstruction.

(a)　　　　　　　　　　　　　　　(b)

Fig. 1. Unknown regions due to reconstruction on a finite domain. A photograph of our "bench" scene is shown in (a), with the reconstruction volume superimposed. Only voxels within the reconstruction volume are considered in voxel coloring algorithms. The scene contains many objects outside of the reconstruction volume that are not reconstructed, resulting in unknown regions that appear as black in a projection of the reconstruction, shown in (b). The ideas presented in this paper warp the voxel space, so that the reconstruction volume can become infinite, and the background scene and environment can be reconstructed.

To address these issues, we propose a warping of the voxel space so that surfaces farther away from the cameras can be modeled without an excessive number of voxels. In addition, our proposed warping of the voxel space can extend to infinity along any dimension, so that infinite (all of R^3), or semi-infinite (such as a hemisphere with infinite radius) reconstruction volumes can be defined. The latter might best model an outdoor scene. As will be shown in subsequent sections of this paper, we develop a hybrid voxel space consisting of an interior space in which voxels are not warped, and an exterior space in which voxels are warped. The voxels are warped so that the following criteria are met:

1. No warped voxels overlap.
2. No gaps form between warped voxels.
3. The warped reconstruction volume is at least semi-infinite.

A voxel coloring algorithm is then executed using the warped reconstruction volume.

The layout of this paper is as follows. First, we explore some related work. Then, we introduce a function that warps the voxel space subject to the criteria enumerated above. Next, we discuss some implementation details that arise when performing a reconstruction in warped space. We then present results that demonstrate the effectiveness of our approach.

2 Related Work

The work presented in this paper is an extension to recent volumetric solutions to the three-dimensional scene reconstruction problem. Seitz and Dyer's [7] voxel coloring technique exploits color correlation of surfaces to find a set of voxels that are consistent with the photographs taken of a scene. Kutulakos and Seitz [5] develop a space carving method that extends voxel coloring to support arbitrary camera placement via a multi-sweep algorithm. Culbertson, Malzbender, and Slabaugh [2] present two generalized voxel coloring (GVC) algorithms, which, like [5] allow for arbitrary camera placement, and in addition use the exact visibility of the scene when determining if a voxel is consistent with the photographs. These three methods, referred to collectively as "voxel coloring algorithms", have been quite successful in reconstructing three-dimensional scenes on a finite spatial domain. In this paper, we extend these three methods in order to reconstruct scenes on an infinite or semi-infinite domain by warping the voxel space used in the reconstruction. Doing so enables the reconstruction of nearby objects, far-away objects, and everything in between.

Saito and Kanade [6], and later Kimura, Saito, and Kanade [4] specify a voxel space using the epipolar geometry relating two [6] or three [4] basis views, for volumetric reconstruction using weakly calibrated cameras. In their approach, a voxel takes on an arbitrary hexahedral shape, a consequence of their projective space. In our approach, we intentionally warp exterior voxels into arbitrarily shaped hexahedra. In [6] and [4], a voxel's size is solely based on its location relative to the cameras that form the basis. In our approach, a voxel's size is instead based on its location in a user-defined voxel space. In [6] and [4], the reconstruction volume is finite, and only foreground surfaces are reconstructed. In contrast, our method warps the voxel space to infinity so that objects far from the cameras can be reconstructed, in addition to foreground surfaces.

In the computer graphics domain, infinite scenes have been modeled and rendered using environment mapping. This method projects the background onto the interior of a sphere or cube that surrounds the foreground scene. Blinn and Newell [1] use such a technique to synthesize reflections of the environment off of shiny foreground surfaces, a procedure also known as reflection mapping. Greene [3] additionally renders the environment map directly to generate views of the background. This approach is quite effective at producing convincing synthetic images. However, since the foreground and background are modeled differently, separate mechanisms must be provided to create and render each. Furthermore, the three-dimensionality of the environment is lost, as the background is represented as a texture-map. Like environment mapping, the techniques described in this paper seek an efficient mechanism to represent the background scene. Our warped volumetric space provides this in a single framework that can more easily accommodate surfaces that appear both in the foreground and background. In addition, we reconstruct the background scene three-dimensionally using computer vision methods.

3 Volumetric Warping

The goal of a volumetric warping function is to represent an infinite or semi-infinite volume with a finite number of voxels, while satisfying the requirement that no voxels overlap and no gaps exist between voxels. There are many possible ways to achieve this goal. In this section, we use the term *pre-warped* to refer to the volume before the volumetric warping function is applied.

The volumetric warping method presented here separates the voxel space into an interior space used to model foreground surfaces, and an exterior space used to model background surfaces, as shown in Figure 2 (a). The volumetric warp does not affect the voxels in the interior space, providing backward compatibility with previous voxel coloring algorithms, and allowing reconstruction of objects in the foreground at a fixed voxel resolution.

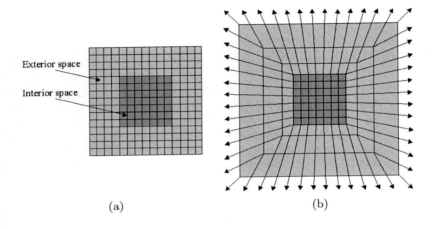

Fig. 2. Pre-warped (a) and warped (b) voxel spaces shown in two dimensions. In (a), the voxel space is divided into two regions; an interior space shown with dark gray voxels, and an exterior space shown with light gray voxels. Both regions consist of voxels of uniform size. The warped voxel space is shown in (b). The warping does not affect the voxels in the interior space, while the voxels in the exterior space increase in size further from the interior space. The outer shell of voxels in (b) are warped to infinity, and are represented with arrows in the figure.

Voxels in the exterior space are warped according to a warping function that changes the size of the voxel based on its distance from the interior space. The further a voxel in the exterior space is located from the interior space, the larger its size, as shown in Figure 2 (b). Voxels on the outer shell of the exterior space have coordinates warped to infinity, and have infinite volume. Note that while the voxels in the warped space have a variable size, the voxel space still has a regular 3D lattice topology.

To help further limit the class of possible warping functions, we introduce the following desirable property of a warped voxel space:

Constant footprint property: For each image, voxels project to the same number of pixels, independent of depth.

Figure 3 shows an example of a voxel space that satisfies the constant footprint property for two cameras. Assuming perspective projection, a voxel space that satisfies this property has a spatially adaptive voxel size that increases away from the cameras, in a manner perfectly matched with the images. While a useful conceptual construct, the constant footprint property cannot in general be satisfied when more than n cameras are present in R^n space. Thus, for three-dimensional scenes, a voxel space cannot be constructed that satisfies the property for general camera placement when there are more than three cameras. Since reconstruction using three or less cameras is limiting, we instead design our volumetric warping function to approximate the constant footprint property for an arbitrary number of images.

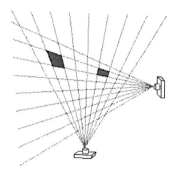

Fig. 3. Example of a 2D voxel space that satisfies the constant footprint property for two images. Notice that the two filled in voxels project to the same number of pixels in the right image, regardless of their respective distance from the camera. Note that this figure is solely used to illustrate the constant footprint property; the warped voxel space developed and used in this paper actually looks like that of Figure 2 (b).

3.1 Frustum Warp

In this subsection, we describe a frustum warp function that is used to warp the exterior space. We develop the equations and figures in two dimensions for simplicity; the idea easily extends to three dimensions.

The frustum warp assumes that both the interior space and the pre-warped exterior space have rectangular shaped outer boundaries, as shown in Figure 4.

The pre-warped exterior space is divided into four trapezoidal regions, bounded by (1) lines l connecting the four corners of the interior space to their respective corners of the exterior pre-warped space, (2) the boundary of the interior space, and (3) the boundary of the pre-warped exterior space. We denote these trapezoidal regions as $\pm x$, and $\pm y$, based on the region's relative position to center of the interior space. These regions are also shown in Figure 4.

Let (x, y) be a pre-warped point in the exterior space, and let (x_w, y_w) be the point after warping. To warp (x, y), we first apply a warping function based on the region in which the point is located. This warping function is applied only to one coordinate of (x, y). For example, suppose that the point is located in the $+x$ region, as depicted in Figure 5. Points in the $+x$ and $-x$ regions are warped using the x-warping function,

$$x_w = x \frac{x_e - x_i}{x_e - |x|}, \qquad (1)$$

where x_e is the distance along the x-axis from the center of the interior space to the outer boundary of the exterior space, and x_i is the distance along the x-axis from the center of the interior space to the outer boundary of the interior space, shown in (a) of Figure 5. A quick inspection of this warping equation reveals its behavior. For a point on the boundary of the interior space, $x = x_i$, and thus $x_w = x_i$, so the point does not move. However, points outside of the boundary get warped according to their proximity to the boundary of the exterior space. For a point on the boundary of the exterior space, $x = x_e$, and so $x_w = \infty$.

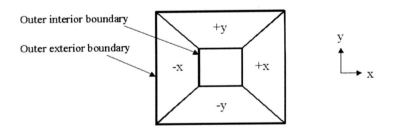

Fig. 4. Boundaries and regions. The outer boundaries of both the interior and exterior space are shown in the figure. The four trapezoidal regions, $\pm x$ and $\pm y$ are also shown.

Continuing with the above example, once x_w is computed, we find the other coordinate y_w by solving a line equation,

$$y_w = y + m(x_w - x), \qquad (2)$$

where m is the slope of the line connecting the point (x, y) with the point a, shown in (b) of Figure 5. Point a is located at the intersection of the line parallel to the x-axis and running through the center of the interior space, with the nearest line l, as shown in the figure. Note that in general, point a is not equal to the center of the interior space.

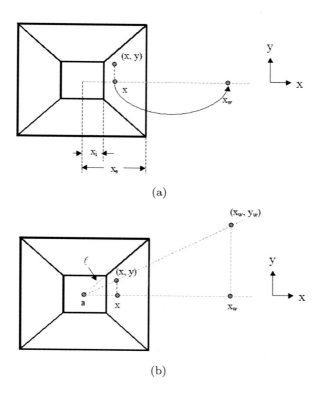

Fig. 5. Finding the warped point. The x-warping function is applied to the x-coordinate of the point (x, y), as the point is located in the $+x$ region. This yields the coordinate x_w, shown in (a). In (b), the other coordinate y_w is found by solving the line equation using the coordinate x_w found in (a).

As shown above, the exterior space is divided into four trapezoidal regions for the two-dimensional case. In three dimensions, this generalizes to six frustum-shaped regions, $\pm x$, $\pm y$, $\pm z$; hence the term *frustum warp*. There are three warping functions, namely the x-warping function as given above, and y- and z-warping functions,

$$y_w = y \frac{y_e - y_i}{y_e - |y|} \tag{3}$$

$$z_w = z \frac{z_e - z_i}{z_e - |z|}, \tag{4}$$

In general, the procedure to warp a point in the pre-warped exterior space is as follows.

1. Determine in which frustum-shaped region the point is located.
2. Apply the appropriate warping function to one of the coordinates. If the point is the in $\pm x$ region, apply the x-warping function, if the point is in the $\pm y$ region, apply the y-warping function, and if the point is the $\pm z$ region, apply the z-warping function.
3. Find the other two coordinates by solving line equations using the warped coordinate.

After reconstruction, we intend the model to be viewed from near or within the interior space. For such viewpoints, voxels will project to approximately the same footprint in each image.

3.2 Other Warping Functions

The frustum warp presented above is not the only possible warp. Any warp that does not move the outer boundary of the interior space, and warps the outer boundary of the pre-warped exterior space to infinity, while satisfying the criteria that no gaps form between voxels, and that no voxels overlap, is valid. Furthermore, it is desirable to choose a warping function that approximates the constant footprint property for the cameras used in the reconstruction as well as the camera placements during new view synthesis. An example of an alternative warping function is one that warps radially with distance from the center of the reconstruction volume.

4 Implementation Issues

Reconstructing a scene using a warped reconstruction volume poses some new challenges, described in this section.

4.1 Cameras Inside Volume

Perhaps the most difficult challenge is that of having the cameras embedded inside the reconstruction volume. Typically, when one uses a standard voxel coloring algorithm, the cameras used to take the photographs of the scene are placed outside of the reconstruction volume, so that at least two cameras have visibility of each voxel. The photo-consistency measure used in voxel coloring algorithms, qualitatively, determines if all the cameras that can see a voxel agree on its color. This photo-consistency is poorly defined when a voxel is visible from only one camera.

Since the warped reconstruction volume can occupy all space, cameras get embedded inside the voxel space, as shown in (a) of Figure 6. Our reconstruction algorithm initially assumes that all voxels are opaque. Therefore, camera views are obscured, and the cameras cannot work together to carve the volume. This poses a problem, since to be properly defined, the photo-consistency measure requires that at least two cameras have visibility of a voxel. Consequently, the voxel coloring algorithm cannot proceed, and terminates without removing any voxels from the volume.

To address this issue, we must remove (pre-carve) a section of the voxel space so that initially, each surface voxel is observed by at least two cameras, validating the photo-consistency measure, as shown in (b) of Figure 6. There are a variety of possible methods to achieve this result. A generic method is to have a user identify regions of the voxel space to pre-carve. Obviously, the pre-carved regions must only consist of empty space, i.e. not contain any scene surfaces to be reconstructed. While effective, this method precludes a fully automatic reconstruction. Alternatively, one can pre-carve the volume using a heuristic. For example, if appropriate, one could require that the cameras have visibility of the boundary between the interior space and the exterior space. Other heuristics are possible. Once the pre-carving is complete, we execute a standard voxel coloring algorithm using the warped voxel space.

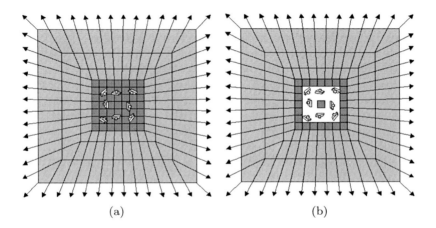

Fig. 6. Pre-carving operation. Reconstruction in the warped space causes the cameras to be embedded in the voxel space, as shown in (a). For many camera placements, it would be impossible to carve any voxels, since no voxel is visible to more than one camera. We execute a pre-carving step in (b) so that cameras can work together to carve the volume.

4.2 Preventing Visible Holes in the Outer Shell

Due to errors in camera calibration, image noise, inaccurate color threshold etc., voxel coloring sometimes removes voxels that should remain in the volume. Thus, it is possible that voxels on the outer shell of the voxel space will be deemed inconsistent. Removing such voxels can result in unknown black regions similar to those in Figure 1 during new view synthesis, as no voxel would project onto the camera for some pixels in the image plane. Since one cannot see beyond infinity, we do not carve voxels on the outer shell of the voxel space, independent of the photo-consistency measure.

5 Results

We have modified the GVC and GVC-LDI algorithms [2] to utilize the warped voxel space. We created a synthetic data set, called "marbles", consisting of twelve 320 x 240 images of five small texture mapped spheres inside a much larger sphere textured with a rainbow-like image. We reconstructed the scene using a voxel space that consisted of 48 x 48 x 48 voxels, of which the inner 32 x 32 x 32 were in the interior space and unwarped. The voxel space was set up so that the five small texture mapped spheres were reconstructed in the interior space, while the larger sphere, making up the background, was reconstructed in the exterior warped space. Sample images from the data set are shown in (a) and (b) of Figure 7. A reconstruction was performed using the warped voxel space. The reconstruction was projected to the viewpoints of (a) and (b), yielding (c) and (d). Note that the background environment was reconstructed using our warped voxel space.

Next, we took a series of ten panoramic (360 degree field of view) photographs of a quadrangle at Stanford University, using a Panoscan[1] digital camera. These photographs had resolution of about 2502 x 884 pixels. One photograph from the set is shown in Figure 8 (a). We have found that when reconstructing an environment, it is preferable to use large field of view images, as objects far from the cameras are visible in many photographs. This achieves a sufficient sampling of the scene with fewer photographs. A voxel space of resolution 300 x 300 x 200 voxels, of which the inner 200 x 200 x 100 were interior voxels, was pre-carved manually by removing part of the voxel space that containing the cameras. Then, the GVC algorithm was used to reconstruct the scene. Figure 8 (b) shows the reconstructed model reprojected to the same viewpoint as in (a). Note that objects far away from the cameras, such as many of the buildings and trees, have been accurately reconstructed. New synthesized views are shown in (c) and (d) of the figure.

Despite the successes of this reconstruction, it is not perfect. The sky is very far away from the cameras (for practical purposes, at infinity), and should therefore be represented with voxels on the outer shell of the voxel space. However, since the sky is nearly textureless, cusping [7] occurs, resulting in inaccurate computed geometry, apparent in an animated sequence of new views of the reconstruction. Reconstruction of outdoor scenes is challenging, as surfaces often do not satisfy the Lambertian assumption. To compensate, we used a higher consistency threshold [7], also resulting in some inaccurate geometry. On the whole,

---

[1] www.panoscan.com

though, the reconstruction is reasonably accurate and produces convincing new views[2].

6 Conclusion

In this paper we have proposed extensions to voxel coloring that permit reconstruction of a scene using a warped voxel space, in an effort to comprehensively reconstruct objects both near and far away from the cameras used to photograph the scene. We have presented a frustum warp function, which describes a method to warp the voxel space to model infinite volumes while maintaining the requirements that no voxels overlap and no gaps form between the warped voxels. We have presented results showing the ability of this approach to reconstruct a background environment, in addition to a foreground scene.

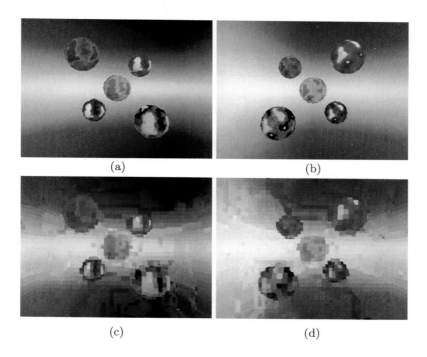

Fig. 7. Original images of the marbles data set are shown in (a) and (b), and a reconstruction projected to the same viewpoints of (a) and (b) is shown in (c) and (d), respectively.

---

[2] An animation showing new synthesized views of our Stanford scene is available online at www.ece.gatech.edu/users/slabaugh/projects/warp.

Fig. 8. Results for the Stanford scene. One of the ten panoramic photographs is shown in (a). The reconstructed model, projected to the same viewpoint as that of (a) is shown in (b). New synthesized panoramic views are shown in (c) and (d).

7 Future Work

Since voxels can warp to points infinitely far from the camera centers, using z-values (such as in a z-buffer) to establish depth order can be problematic due to a computer's finite precision. We are interested in exploring alternate methods, such as painter's algorithms, to determine depth order of voxels during reconstruction and rendering.

Acknowledgments. We are indebted to Panoscan for acquisition of the panoramic images of Stanford's campus, and to Mark Livingston and Irwin Sobel for calibration of these images. We further thank Mark Livingston for developing a viewer that renders a reconstructed scene using volumetric warping. We would also like to thank Steve Seitz and Chuck Dyer for numerous discussions regarding voxel coloring. Finally, we thank Fred Kitson and Ron Schafer for their continued support and encouragement of this research.

References

1. J. Blinn and M. Newell, "Texture and Reflection on Computer Generated Images," *Communications of ACM*, Vol. 19, No. 10, pp. 542 - 547, Oct. 1976.
2. W. B. Culbertson, T. Malzbender, and G. Slabaugh, "Generalized Voxel Coloring," *Vision Algorithms: Theory and Practice*, Springer-Verlag Lecture Notes in Computer Science, pp. 100 - 115, 1999.
3. N. Greene, "Environment Mapping and Other Applications of World Projections," *IEEE Computer Graphics and Applications*, pp. 21 - 29, Nov. 1986.
4. M. Kimura, H. Satio, and T. Kanade, "3D Voxel Construction Based on Epipolar Geometry," *Proc. ICIP*, pp. 135 - 139, 1999.
5. K. Kutulakos and S. Seitz, "A Theory of Shape by Space Carving," *Proc. ICCV*, pp. 307 - 314, 1999.
6. H. Saito and T. Kanade, "Shape Reconstruction in Projective Grid Space from Large Number of Images," *Proc. CVPR*, pp. 49 - 54, 1999.
7. S. Seitz and C. Dyer, "Photorealistic Scene Reconstruction by Voxel Coloring," *International Journal of Computer Vision*, Vol. 35, No. 2, pp. 151-173, Feb. 1999.

Discussion

1. **David Nister, Ericsson**: If you know the cameras and the resolution of all the images, you can determine for every point in space what intrinsic resolution you have there. I was wondering if you could comment on if your warping function corresponds to that?
   Gregory Slabaugh: If one did such an analysis, at each point in space one would find a different intrinsic spatial resolution resulting from each camera. In general, there is no voxel space that perfectly matches the intrinsic resolution (i.e. satisfies the constant footprint property discussed in our paper, with a footprint of one pixel) for all cameras, when the number of cameras

is greater than three. So instead, our warping function approximates this property, by requiring that an exterior voxel's size increases linearly with distance from the interior region. Thus, exterior voxels will project to approximately the same number of pixels in an image, regardless of the voxel's distance from the camera. In order for this to work properly, it is necessary that both the reference and the virtual viewpoints are in or near the interior region of the warped voxel space.

2. **Bill Triggs, INRIA Rhône-Alpes**: Two questions. Firstly, how do you decide how many voxels to devote to modelling the exterior?

   Gregory Slabaugh: That is a good question. How far you move in each slice is a function of how many voxels are used to model the scene. For example, in the Stanford data set we have $300 \times 300 \times 200$ voxels of which the inner $200 \times 200 \times 100$ are interior. So we set up the voxel space so that in each direction there are 100 voxels in the exterior region, 50 of which are on either side of the reconstruction volume. Now if you have more voxels in this exterior space you are going to get a better resolution as you go out to infinity.

3. **Bill Triggs**: Secondly, you said that you have problems with Z-buffer resolution for distant points. Would it be possible to prewarp the depths in some way to avoid this problem?

   Gregory Slabaugh: That's a great observation. We have looked into that a little bit and we are still working on it.

4. **Paul Debevec, University of Southern California (comment)**: Specifically on the Z-buffer, I know that like in OpenGL the Z-buffer is in projective coordinates anyway. So you can have things go out to infinity. The problem is you have a near clipping plane and a far clipping plane. There is no problem with putting the far clipping plane at infinity as long as the near clipping plane isn't at zero.

   Gregory Slabaugh: We have done all our rendering in software we coded ourselves, so maybe we should take a look at OpenGL. The problem is that of representing a huge dynamic range of Z values with a finite precision (32-bit) Z-buffer; there isn't enough resolution.

5. **Kyros Kutulakos, University of Rochester**: You suggested there are many different warping functions, one of which involves choosing a warping function where the number of pixels contained in a voxel projection is constant. I wonder why you didn't choose that particular warping function given that when you don't obey that particular constraint, if you move farther and farther away, your voxels will project onto more and more pixels. Making it more difficult to establish consistency given that there going to be different colours and intensities that lie inside the voxel projections. And in relation to that could you say a little bit about what thresholds you use for this particular scene?

   Gregory Slabaugh: I don't have the numbers for the thresholds right off the top of my head and they probably wouldn't be too interesting. But what we saw when doing our consistency measure, we take a voxel and project it into all the images that can see that voxel and collect the pixels in these views. Then what we do, just as in some of your work, is we take the mean and the standard deviation and threshold the standard deviation. If a voxel is large in one particular image, it can bias the consistency measure. To avoid this, we want the voxels to project onto approximately the same number of pixels. In general, no voxel space exists for which voxel footprints are constant when more than three cameras are used to photograph the scene. So instead our warping function tries to approximate this constant footprint property, for viewpoints in or near the interior region. Thus, we require that the cameras used to photograph the scene, and the new synthesized viewpoints, are located in or near the interior region.

6. **Andrew Fitzgibbon, University of Oxford**: This may be naive, I have never tried to code one of these things, but if you used something like an octree, then it would be easy to devise a partitioning strategy that starts infinitely large and adapts its resolution. Is that not an option?

   Gregory Slabaugh: Yes, that is a great question. That is certainly a possibility that we haven't implemented. Andrew Prock at Wisconsin has done some great work for hierarchical voxel colouring and we are interested in using their techniques and adapting them to ours. I think that could be fruitful.

7. **Michal Irani, The Weizmann Institute of Science**: When you showed the video sequence at the end, there seems to be some non-rigidity. I was wondering if it was 3D but non-rigid, or if it was a particular artifact of this warping technique that made it look this way, or if it is a problem with the epipolar constraint estimation.

   Gregory Slabaugh: When we reconstructed this scene we used panoramic images and we re-rendered the reconstruction using a panoramic transform as well. So you will notice that objects at the right edge of the image loop around to the left edge. So that might be producing some of the effects that you're describing.

8. **Marc Pollefeys, K.U.Leuven**: Would it be possible to extend this representation so that it allows walk-through applications? In this case one would probably have to be able to switch between different models.

   Gregory Slabaugh: One weakness of our approach, at least in the way we presented it here, is that we have just one interior space. To do what you're describing, we might want to have multiple interior spaces and a way to combine warped voxel spaces together, this being interesting future research.

A Compact Model for Viewpoint Dependent Texture Synthesis

Alexey Zalesny[1] and Luc Van Gool[1,2]

[1]D-ELEK/IKT, ETH Zurich, Switzerland
{zalesny, vangool}@vision.ee.ethz.ch
www.vision.ee.ethz.ch/~zales

[2]ESAT-PSI, Kath. Univ. Leuven, Belgium
Luc.VanGool@esat.kuleuven.ac.be

Abstract. A texture synthesis method is presented that generates similar texture from an example image. It is based on the emulation of simple but rather carefully chosen image intensity statistics. The resulting texture models are compact and no longer require the example image from which they were derived. They make explicit some structural aspects of the textures and the modeling allows knitting together different textures with convincingly looking transition zones. As textures are seldom flat, it is important to also model 3D effects when textures change under changing viewpoint. The simulation of such changes is supported by the model, assuming examples for the different viewpoints are given.

1 Introduction

Increasingly, the computer vision and graphics communities turn toward the 3D reconstruction of large scenes. Not all parts of such scenes are equally interesting. An architectural highlight like a monument may be surrounded by streets with hundreds of normal houses. An archaeological site may contain interesting ruins that are dispersed in the landscape. Realistic visualization nevertheless imposes that the "less interesting" parts are displayed at the same resolution as the interesting ones. The synthesis of realistic textures can be part of the solution. Brick walls, grass, rocks, sand, concrete, vegetation, ... can be emulated based on a compact model of these textures.

Several powerful texture synthesis methods have been proposed over the last couple of years [3, 4, 7, 11, 15, 19]. The realism of synthesized textures has gone up dramatically. With this paper we hope to contribute in a number of respects:

- *The texture* models *are very compact, yielding excellent compression.* In contrast to several recent methods, the model doesn't contain an example image of the texture.
- *No verbatim repetitions of parts in stochastic textures.* There is no copying of patterns from the example image involved.
- *Perceptually convincing transitions where textures meet.* Seams between similar or

different textures can be eliminated through the similar procedures as those used for texture synthesis.

- *Fast and compact inclusion of 3D effects.* The very existence of texture is usually due to the fact that the surface is not really flat. Hence, changing viewpoint entails more than simple foreshortening, although this is common practice in texture mapping. Effects like self-occlusion and different changes in the angle between the normals and the viewing directions are not taken into account through foreshortening. Our model can be adapted quickly to include these effects.

2 Clique Selection

Our approach extracts statistical properties from an example texture, which are then combined into a texture model. From this model more of the same texture is generated, i.e., textures that have similar statistics. Such texture synthesis methods differ in the properties that they extract and the algorithms to generate images with the prescribed statistics. The following sections describe these aspects for our approach.

2.1 Extracted Statistical Properties

The method extracts only first- and second-order statistics. This is in line with Julesz's observation that first and second-order statistics govern to a large extent our perception of textures. Yet, Julesz also demonstrated that third and higher order statistics couldn't be neglected just like that, mainly because of figural patterns that are not preserved [12]. As we will demonstrate, quite a broad range of textures can be synthesized nevertheless and in fact higher-order statistics can be included in the model, at the expense of computation time.

The first order statistics are characterized through the intensity histogram $f(q)$, where q is intensity.

The second-order statistics draw upon the cooccurrence principle: for point pairs at fixed relative positions the intensities are compared. The point pairs are called *cliques* and pairs with the same relative positions (translation invariance) form a *clique type* (Fig. 1). Individual cliques of type α will be denoted as κ_α.

Cliques of the same type   Cliques of different types

Fig. 1.   Cliques and clique types.

The cliques are ordered sets. Hence, a "tail" and "head" pixel can be distinguished. Instead of storing the complete joint probability distributions for the different clique types, our model only stores the distribution of the intensity differences between the head and tail pixels. The original intensities are requantized into 32 levels, leading to 63 signed difference values. For a clique of type α the distribution of these signed difference values Δ is denoted as $f_\alpha(\Delta)$.

The texture model consists of two parts. A first part specifies the clique types that are used to describe the texture. Including all possible clique types would make the texture model prohibitively large, hence, a limited number of them will be selected. The set of these clique types is called the *neighborhood system*. A second part is the *statistical parameter set*: the distributions $f(q)$ and $f_\alpha(\Delta)$ for the selected clique types. The next section proposes a strategy to select only a few clique types, but with maximal effect.

2.2 Clique Type Selection

Textures are synthesized by mimicking the statistics of the example texture for the different clique types. As including all clique types in the model is not a viable option, a good selection needs to be made. One criterion is to consider all clique types up to a maximum head-tail length [5]. The maximum length is then quite low by necessity, excluding longer-range interactions. We put this maximum rather high (45) but only select a subset of the corresponding clique types. The selection is based on their impact on the target statistics as explained next. Clique types are added one by one to the model, through the following algorithm:

step 1. Collect the complete 2nd-order statistics for the example texture, i.e., the intensity difference distributions of all clique types up to a maximum length. After this step the example texture is no longer needed.

step 2. Generate an image filled with independent noise with values uniformly distributed in the range of the example texture. This noise image serves as the initial synthesized texture, to be refined in subsequent steps.

step 3. Collect the pairwise statistics of all clique types (up to the same maximal length) for the current synthesized image (initially noise).

step 4. For each clique type, compare the difference distributions of the example texture and the synthesized texture by calculating the Euclidean distance.

step 5. Select the clique type with the maximal distance. If this distance is less than some threshold go to step 8 – the end of the algorithm. Otherwise, add the clique type to the (initially empty) neighborhood system and its difference distribution to the (initially empty) parameter set.

step 6. Synthesize a new texture using the updated neighborhood system and parameter set. The texture should have the prescribed statistics for all clique types in the neighborhood system.

step 7. Go to step 3.

step 8. End of the algorithm.

The distribution distances that are compared between clique types in step 4 are weighted with the number of cliques. This should prevent unstable statistical behavior when there are only few cliques (typical for long clique types).

Gimel'farb [6] uses a similar approach, but selects all clique types simultaneously and independently. To get at the same quality of the synthesized textures about 5 times as many clique types need to be included. Texture analysis is faster but — and this is more critical — texture synthesis is about 5 times slower.

For this texture analysis algorithm, repeated texture synthesis is necessary (step 6). We use the same algorithm as for the synthesis from the final texture model. This algorithm is described in the next section. First, the above analysis algorithm is illustrated and its extension to color images is discussed.

Fig. 2 left shows an example texture. A model of this texture has been built. Fig. 2 right shows the synthesis result. The left column in Fig. 3 shows a series of intermediate, synthesized textures as new clique types are added to the neighborhood system, shown in the right column.

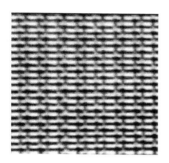 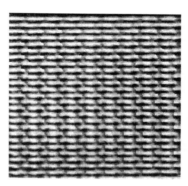

Fig. 2. Left: an example texture (straw cloth, Brodatz D53) that is to be modeled. Right: final synthesis.

The neighborhood systems show, which cliques the central pixel is a member of. Note that every clique type adds two such cliques: the central pixel can play the role of both head and tail, hence, the point symmetry. In these schematic drawings of the neighborhood systems one also notices that the central pixel itself is included. This is to indicate that also first-order statistics about the intensity of individual pixels is part of the statistical parameter set.

In the case of color images, separate neighborhood systems are selected for each of the three color bands. Besides these within-band statistics, pairwise interactions between the color bands have to be included. An example of such within-band + inter-band neighborhood system is shown in Fig. 4. The three (R, G, and B) intensity histograms and some interactions are always included into the neighborhood system. These are the interactions with the four nearest neighbors within the bands and the "vertical" connections between bands (i.e., between identical pixels). Experiments have shown that they had to be included almost without exception in the different texture models. Their automatic inclusion helps to speed up the modeling.

After the 8-step texture modeling algorithm we have the final neighborhood system of the texture and its parameter set. This model is very compact compared to

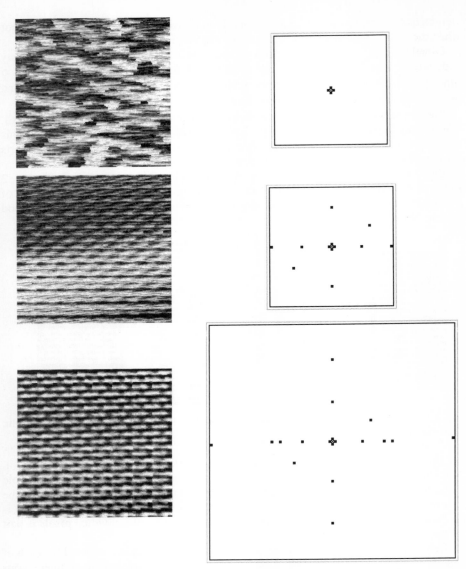

Fig. 3. Left: subsequent synthesized textures for the example texture in Fig. 2. Right: selected clique types after 2, 6, and 9 analysis iterations.

the complete 2nd-order statistics extracted in step 1. Typically, only 10 to 40 clique types (20 to 80 neighbors of a pixel) are included. The model size amounts to a few hundred or maximally a few thousand bytes. Nevertheless, the differences between synthesized and example statistics are very small for all clique types, including the ones that have not been selected.

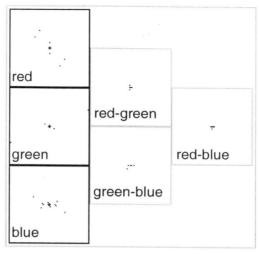

Fig. 4. Complete neighborhood system for a color texture. Left column: separate neighborhood systems for the 3 color bands Red, Green, and Blue. Second and third columns: neighborhood systems for pairwise interactions between the R-G, G-B, and R-B bands.

2.3 Texture Synthesis

For texture synthesis, the images are treated as a realization from the family of Markov random fields with the extracted neighborhood system. For notational simplicity we will drop 1st order statistics terms from the formulas in the sequel. The intensity histograms are used in exactly the same way as the second order difference distributions and the corresponding cliques can be thought to collapse to a single pixel.

The synthesis proceeds iteratively to obtain the same parameter set as the example texture. To that end, the Gibbs potentials in the exponential representation of the field's probability distribution are iteratively updated. The joint probability $\mathbf{P}(s)$ of an image s is expressed as

$$\mathbf{P}(s) \propto \exp \sum_{\alpha} \sum_{\kappa_\alpha} g_{\alpha,\Delta} = \exp \sum_{\alpha} \sum_{\Delta} n_\alpha(\Delta) g_{\alpha,\Delta} \qquad (1)$$

with $g_{\alpha,\Delta}$ the Gibbs potential for the clique type α and the intensity difference Δ, and $n_\alpha(\Delta)$ the number of cliques of clique type α with intensity difference Δ. The double sum adds the potentials for all cliques of the different clique types present in the neighborhood system.

The Gibbs potentials are real numbers that have to be manipulated in order to approach the target statistics. This iterative process has two components: 1) modify the texture synthesized so far based on the latest potentials, and 2) update the potentials according to the deviations of the modified texture from the target.

For the first part, pixels are selected randomly. The Metropolis stochastic relaxation procedure is used [14] to update the intensity value of a selected pixel. Given the neighborhood system and the current Gibbs potentials the probability of having intensity s_i at a pixel i is given by the single-point Markov conditional probability

$$p(s_i \mid neighbors) = \frac{\exp \sum_\alpha \sum_{\kappa_\alpha \ni i} g_{\alpha, \Delta(\kappa_\alpha)}}{\sum_{s_i'} \exp \sum_\alpha \sum_{\kappa_\alpha \ni i} g_{\alpha, \Delta'(\kappa_\alpha)}} \qquad (2)$$

where $\{\kappa_\alpha : i \in \kappa_\alpha\}$ are the cliques of type α that contain pixel i (usually 2, once as head and once as tail) and where $\Delta'(\kappa_\alpha)$ each time denotes the signal difference corresponding to an intensity s_i' at the pixel position. The new signal level is selected uniform-randomly from the given range. Then according to the transition probability (2) the Metropolis updating rule is as follows:

$$p(new_i) = \begin{cases} 1, & p(new_i \mid neighbors) \geq p(old_i \mid neighbors), \\ \dfrac{p(new_i \mid neighbors)}{p(old_i \mid neighbors)}, & otherwise. \end{cases} \qquad (3)$$

Within one "Metropolis iteration" all points are visited once and updated in this way.

Then the new statistics are derived and the Gibbs potentials are updated as:

$$\forall \alpha, \Delta \quad g_{\alpha,\Delta}^{\tau+1} = g_{\alpha,\Delta}^{\tau} + c \left(f_\alpha(\Delta) - f_\alpha^\tau(\Delta) \right) \qquad (4)$$

where τ is the iteration number, c is a small constant, and the expression between parentheses is the difference between the target difference distribution and the one realized at iteration τ for the difference Δ and clique type α. The Gibbs potential and hence, the probabilities for a specific Δ and α increase if $f_\alpha^\tau(\Delta)$ is too low. They are seen to decrease in the opposite case, again pushing intensities in the right direction.

The overall synthesis algorithm then goes as follows:

step 1. Put the initial Gibbs potentials to zero, fill the initial synthesized image with white noise.
step 2. Calculate the new statistics and update the Gibbs potentials accordingly.
step 3. Update the image by performing a Metropolis iteration. If the iteration number surpasses a limit, go to step 5.
step 4. Go to step 2.
step 5. End of the algorithm.

The convergence of this relaxation procedure has been proven [18].

This procedure is also used in step 6 of the texture analysis algorithm but the potentials are not reset to zero and the image is not reset to noise on the intermediate analysis stages, which have then about 10 times smaller iteration number.

2.4 Examples of Synthesized Textures

This section shows a few examples, obtained with the proposed Clique Selection Method. Fig. 5 and Fig. 6 show example textures on the left and synthesized textures on the right. On the whole, the synthesized textures are perceptually similar to the examples for both regular and stochastic textures. Nevertheless, some of the examples indicate that our approach finds it difficult to capture complex orderings and the precise shapes of texels. The structure in Fig. 5 d) is not completely preserved and the texels in Fig. 6 d) are deformed. This is a consequence of only considering pairwise pixel interactions.

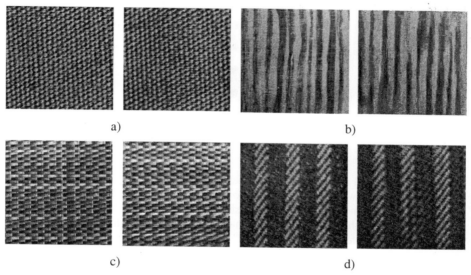

Fig. 5. Left: originals; right: synthetic; a) Brodatz D77, cotton canvas; b) Brodatz D50, raffia woven with cotton threads; c) Brodatz D55, straw matting; d) Brodatz D11, homespun woolen cloth.

Fig. 7 shows two more examples, where the method fails dramatically. Again, this is due to the presence of precisely shaped texels placed irregularly in the first case, and the complex mix of curvilinear and blob-like structures in the second case.

The synthesis of textures is useful when reconstructing large-scale environments. As already mentioned in the introduction, many parts will not have to be modeled in great detail, yet have to be visualized at the same high resolution as the objects of interest. In the context of the European project Murale, we are in the process of building an extensive 3D model of the archaeological site at Sagalassos in Turkey. Ruins are dispersed in a landscape of many squared kilometers. Only the ruins are modeled in detail, the mixture of grass, sand, and rocks in between should look real, without a need to precisely reflect reality. Even if one wanted to model these vast parts based on images, this would take an awful lot of time and memory. Texture synthesis is more viable option. Fig. 8 shows an example image of Sagalassos landscape texture. The figure also shows texture synthesized from this example. This

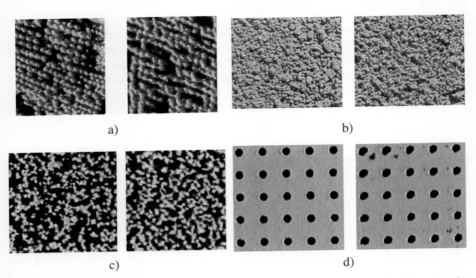

Fig. 6. Left: originals; right: synthetic; a) aerial photograph of forest; b) coffee grounds (MIT VisTex); c) algae (MIT VisTex); d) ceiling tile (MIT VisTex).

texture was modeled and then large patches of similar texture were synthesized. The synthesized texture was then mapped onto the 3D landscape model. Fig. 9 shows a fragment of the 3D model. The top image shows part of a 3D building model, inserted in the 3D model of the landscape, which has much lower resolution, both in terms of the geometry and in terms of the texture. The bottom image shows the result when the synthesized texture is mapped onto the landscape model. The result looks better, although now the coarseness of the geometry becomes more salient. Of course it could be smoothed.

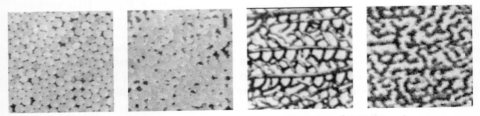

Brodatz D66. Plastic pellets. Brodatz D87. Sea fan.

Fig. 7. Textures that cannot be reproduced with only pairwise interactions.

The example of the archaeological site automatically leads to two further considerations:
texture knitting: natural textures will not be sharply delineated and we need to provide naturally looking transitions between different textures. Also, when mapping synthesized textures onto the 3D model, seams will show up, even between similar textures. These have to be removed.

3D effects: textures are not flat and will often have to be mapped on curved surfaces. Simple foreshortening of a texture will not generate the required 3D effects. The synthetic texture will only look natural from viewpoints similar as that of the example image from which it was generated. Solving this shortcoming calls for models that take 3D effects like viewpoint dependent degrees of self-occlusion and reflectance characteristics into account.

These are the subjects of the next section.

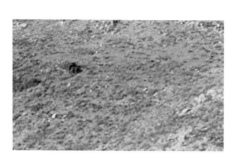

Fig. 8. Left: Image showing terrain texture at the archaeological site of Sagalassos. Right: synthetic texture based on the texture example.

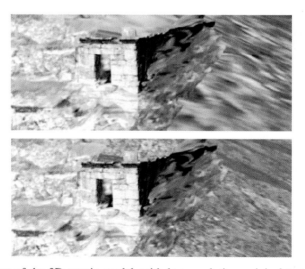

Fig. 9. Top: part of the 3D terrain model, with low-resolution, original texture. Bottom: the same scene with synthetic texture mapped on. The resolution of the landscape texture now better matches that of the building.

3 Texture Knitting and 3D Effects

Often more than one texture has to be mapped onto a surface. In the example of the archaeological site, this could be a texture for rock and a texture for grass. Example textures for both are shown in Fig. 10, on the left. The central area that contains both types of textures was analyzed with the modeling algorithm. Then, the colors of the pixels in a central area (not necessarily the same) were replaced by colors synthesized based on this model. The result is that automatically on the grass side a texture is generated that looks more like grass with a little bit of rock and v.v. This effect is much desirable and follows automatically from the fact that the clique types with shorter lengths prescribe colors more similar to those of the unmixed texture a pixel is closest to.

 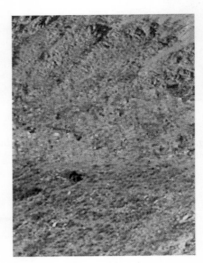

Fig. 10. Left: two example textures (rock and grass). Right: the same two textures, but with a gradual transition inserted for an area around the textures' boundary.

A similar procedure is useful for the seamless knitting of patches of the same texture. Fig. 11 left shows four patches of the same texture. One can see sharp unnatural boundaries between the patches. New texture was synthesized in a region around the boundaries, using the model of this texture. The left figure was used as the initial texture to start the synthesis iterations from. The result is shown in Fig. 11, right. The seams disappear because the newly generated intensities are based on information from either side.

Another issue is the emulation of 3D effects. As most textures are not flat, changing the viewpoint will have a more drastic effect than just a foreshortening of the texture along the direction of the slant [1, 2]. Occlusions are only one example of the phenomena that defy such simple model. In our work we have chosen a simple modification of the texture synthesis algorithm that avoids the need for a complete analysis for all the viewpoints. In particular, we avoid extracting a new neighborhood system for every viewpoint. The texture is modeled for one viewpoint, typically for a

Fig. 11. Left: four patches of the same texture. Right: seamless knitting of patches.

fronto-parallel one. The neighborhood system for that viewpoint is then deformed by contraction or stretching in the direction of the slant change for other viewpoints. This does not provide for the required 3D effects *per se*, but already yields a first approximation that is modified further. These further modifications – necessary to capture 3D effects other than foreshortening - are obtained from the second component of the texture model: the intensity statistics. The intensity histograms and difference distributions for the affinely deformed neighborhood system are learned anew from an example image for the new viewpoint. This process is very fast (milliseconds compared to the tens of minutes required for the extraction of a new neighborhood system). As a consequence, building a texture model that includes 3D effects takes virtually no additional time compared to the extraction of a model for a single viewpoint.

Fig. 12 shows in its top row three original images of the same texture, but viewed from three different angles (Columbia-Utrecht image database, CUReT [2]). Image b) was used to extract a complete texture model, i.e., a neighborhood system and the corresponding difference distributions. Image e) shows a texture that has been synthesized on the basis of this model. The flanking images d) and f) have been created by the method that has just been described. The neighborhood system of the middle image (shown in h)) has been stretched as a first step in the generation of d) and has been contracted for f). These deformed neighborhood systems are shown in g) and i) respectively. Then, from the images a) and c) new intensity statistics (difference distributions and intensity histograms) are extracted. Textures d) and f) have been generated from the deformed neighborhood systems g) and i) in combination with the difference distributions and intensity histograms of a) and c). As can be seen, the similarity is quite good. As is also clear from inspection, an image like a) cannot be produced from b) through simple stretching alone.

0 shows a second example. Image a) shows texture of straw. Image b) shows a result obtained with the texture synthesis algorithm based on a neighborhood system and intensity statistics extracted from a). Image c) shows a real image of the same straw structure for an oblique view. Image d) shows what happens if one would simply contract the image b). As can be seen, this simple procedure leads to strong perceptual differences. Image e) is the result of texture synthesis based on a model extracted from c), i.e., a completely new neighborhood system and its intensity statistics. Image f) finally shows the result of texture synthesis based on a deformed

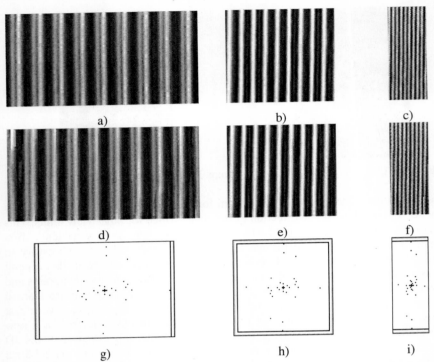

Fig. 12. Three oblique views of a real texture (CUReT); a) original image for a viewing angle 11° away from perpendicular; b) same for 56°; c) same for 79°; e) synthetic texture based on a neighborhood system – h) – and difference distributions learned from b); d) and f) synthetic results based on neighborhood systems g) and i), which are transformed versions of h).

(contracted) version of the neighborhood system of b) combined with intensity statistics for this contracted neighborhood system extracted from image c). This result seems as good as e) but is obtained much faster.

Fig. 14 gives more examples of texture synthesis based on deformed neighborhood systems. The oblique, synthesized texture in every block was synthesized by deforming the neighborhood system of the head-on views (top left) combined with the corresponding intensity statistics as extracted from the original oblique views (top right).

4 Conclusions and Future Work

A texture synthesis method was proposed that builds compact texture models based on 1st and 2nd-order statistics. This method is a further development and specialization of earlier work [8- 10, 17, 18], which partly targeted other applications like texture based segmentation and image retrieval.

Examples show that it is able to produce textures that look very similar to the example textures from which the models are learned and this for a rather broad class of textures. Random field theory is used "only" as a tool for generating the synthesized texture sequences with gradually changing Gibbsian transition probabilities. In this respect, the work is similar to that reported in [16, 19].

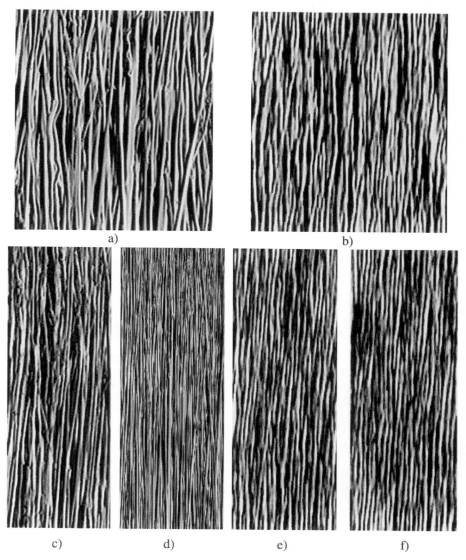

Fig. 13. Straw (CUReT, 40b); a) original image for a perpendicular view; b) synthesized texture based on a model for a); c) original image for an oblique view (68°); d) result of contracting b); e) texture synthesis based on a completely new model extracted from c); f) texture synthesis based on a transformed neighborhood system for b) and new difference distributions from c).

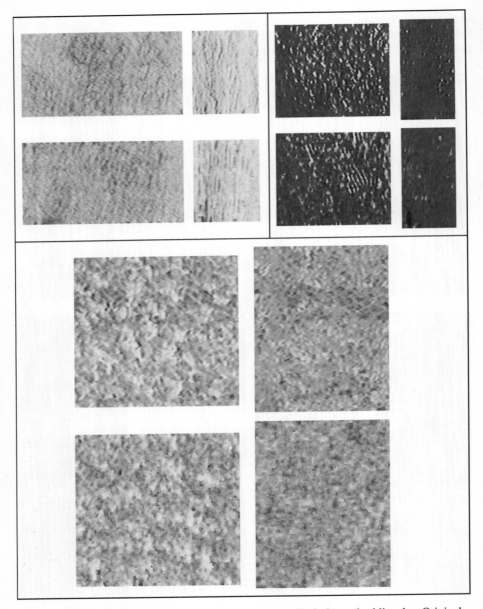

Fig. 14. Different textures (CUReT) viewed perpendicularly and obliquely. Original images in the top rows and synthetic images in the bottom rows of each frame. The bottom right images were synthesized from deformed neighborhoods.

Nice features of the proposed approach are that the original texture is not needed for synthesis and that no disturbing repetitions of patterns occur, even if large areas of synthetic texture are produced. These aspects may be an advantage with respect to the texture synthesis method of De Bonet [3]. The latter also has difficulties when the raster size of the synthesized image is not a multiple of the period of the textural pattern (Fig. 15) or if the main structural elements of the texture are slightly rotated with respect to horizontal and vertical directions, which have a special status for the underlying pyramid used by this method (Fig. 16). The original texture patch was taken from the De Bonet's web site [3], zoomed in to 120% in the first case or rotated to 30 degree in the second case.

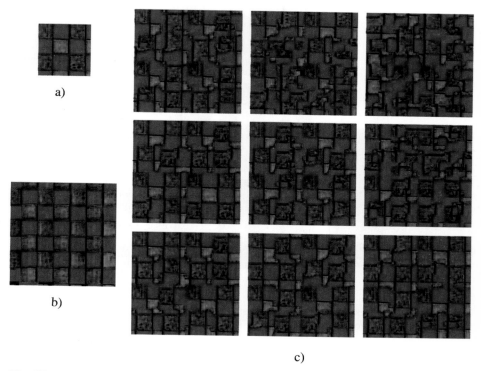

Fig. 15. a) original texture; b) texture synthesized by our method; c) texture synthesized by De Bonet's algorithm for 9 different parameter settings (from high regularity – bottom left – to high randomness – top right).

Limiting the modeling to 1st and 2nd order statistics restricts the class of textures that can be handled successfully. Future work will be aimed to include at least part of the higher order statistics, without increasing computer time too much. Similarly, the second-order statistics are characterized through simple difference distributions. It will be worthwhile to consider more sophisticated features. Also, the current analysis is based on raw intensity data, whereas the responses of filters could be used as input. The promise held by adding filter responses was clearly illustrated by recent work of Leung and Malik [13], who could replicate textures convincingly including 3D effects. Our work differs in that the focus is on the synthesis of new texture rather than precisely replicating textures presented to the system.

Computation time currently is quite high, especially for the texture modeling and may go up to more than one hour CPU time for a 200x200 image on an SGI O_2. Fortunately, the modeling needs to be done only once. Texture synthesis is much faster, but also takes tens of minutes. The typical amount of synthesis iterations lies between 1000 and 3000. Increasing the speed is another topic for future research.

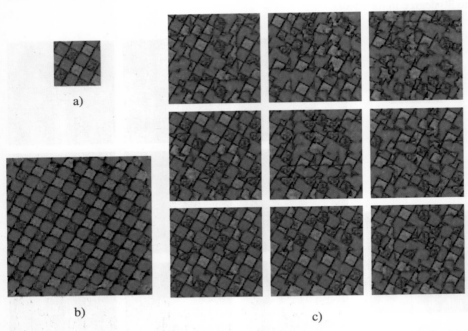

Fig. 16. a) original texture; b) texture synthesized by our method; c) texture synthesized by De Bonet's algorithm for 9 different parameter settings (from high regularity – bottom left – to high randomness – top right).

The paper also presented work to generate natural transitions between different textures and to mimic 3D effects based on a purely 2D representation. The latter work can be further refined, e.g. by investigating into the evolution of the difference distributions in the model with changing viewing angle. It is to be expected that the distributions can be expressed more concisely through principal components. The approach we used is slightly similar to that of Hsu and Wilson [11] who combined affinely distorted texels with statistical variations. However, that work also builds a replica of a given texture, rather than generating new, but similar texture. It also is not based on Gibbs models.

Acknowledgment. The authors gratefully acknowledge support by the European ISP project "MESH".

References

1. K. Dana and S. Nayar. Correlation Model for 3D Texture. *Proc. Int. Conf. Computer Vision (ICCV'99)*, 1999, pp. 1061-1066.
2. K.J. Dana, B. Van Ginneken, S.K. Nayar, and J.J. Koenderink. Reflectance and Texture of Real-World Surfaces. *ACM Transactions on Graphics*, Vol. 18, No. 1, 1999, pp. 1-34.
3. J.S. De Bonet. Multiresolution Sampling Procedure for Analysis and Synthesis of Texture Images. *Proc. Computer Graphics, ACM SIGGRAPH'97*, 1997, pp. 361-368.
4. A. Efros and T. Leung. Texture synthesis by non-parametric sampling. *Proc. Int. Conf. Computer Vision (ICCV'99)*, Vol. 2, 1999, pp. 1033-1038.
5. A. Gagalowicz and S.D. Ma. Sequential Synthesis of Natural Textures. *Computer Vision, Graphics, and Image Processing*, Vol. 30, 1985, pp. 289-315.
6. G. Gimel'farb. On the Maximum Likelihood Potential Estimates for Gibbs Random Field Image Models. *Proc. Int. Conf. Pattern Recognition (ICPR'98)*, Vol. II, 1998, pp. 1598-1600.
7. G. Gimel'farb, *Image Textures and Gibbs Random Fields*. Kluwer Academic Publishers: Dordrecht, 1999, 250 p.
8. G. Gimel'farb and A. Zalesny. Low-Level Bayesian Segmentation of Piecewise-Homogeneous Noisy and Textured Images. *Int. J. of Imaging Systems and Technology*, Vol. 3, No. 3, 1991, pp. 227-243.
9. G.L. Gimel'farb and A.V. Zalesny. Probabilistic Models of Digital Region Maps Based on Markov Random Fields with Short a Long-Range Interaction. *Pattern Recognition Letters*, Vol. 14, Oct. 1993, pp. 789-797.
10. G.L. Gimel'farb and A.V. Zalesny. Markov Random Fields with Short and Long-Range Interaction for Modeling Gray-Scale Texture Images. *Proc. 5th Int. Conf. Computer Analysis of Images and Patterns (CAIP'93)*, Lecture Notes in Computer Science 719, Berlin: Springer-Verlag, 1993, pp. 275-282.
11. T.I. Hsu and R. Wilson. A Two-Component Model of Texture for Analysis and Synthesis. *IEEE Trans. on Image Processing*, Vol. 7, No. 10, Oct. 1998, pp. 1466-1476.
12. B. Julesz and R.A. Schumer. Early visual perception. *Ann. Rev. Psychol.* Vol. 32, 1981, pp. 575-627 (p. 594).
13. T. Leung and J. Malik. Recognizing surfaces using three-dimensional textons. *Proc. Int. Conf. Computer Vision (ICCV'99)*, 1999, pp. 1010-1017.
14. N. Metropolis, A. Rosenbluth, M. Rosenbluth, A. Teller, and E. Teller. Equations of State Calculations by Fast Computing Machines. *J. Chem. Phys.*, Vol. 21, 1953, pp. 1087-1091.
15. J. Portilla and E. P. Simoncelli. Texture Modeling and Synthesis using Joint Statistics of Complex Wavelet Coefficients. *Proc. of the IEEE Workshop on Statistical and Computational Theories of Vision*, Fort Collins, CO, June, 1999, www.cis.ohio-state.edu/~szhu/SCTV99.html.
16. Y.N. Wu, S.C. Zhu, and X. Liu. Equivalence of Julesz and Gibbs texture ensembles. *Proc. Int. Conf. Computer Vision (ICCV'99)*, 1999, pp. 1025-1032.
17. A.V. Zalesny. Homogeneity & Texture. General Approach. *Proc. 12th IAPR Int. Conf. Pattern Recognition (ICPR'94)*, Vol. 1, 1994, pp. 525-527.
18. A. Zalesny. *Analysis and Synthesis of Textures with Pairwise Signal Interactions*. Tech. Report KUL/ESAT/PSI/9902, Katholieke Universiteit Leuven, Belgium, 1999, p. 132, www.vision.ee.ethz.ch/~zales.
19. S.C. Zhu, Y.N. Wu, and D. Mumford. Filters, Random Fields And Maximum Entropy (FRAME). *Int. J. Computer Vision*, Vol. 27, No. 2, March/April 1998, pp. 1-20.

Discussion

1. **Andrew Fitzgibbon, University of Oxford:** To compare your distributions you use the Euclidean distance, does it make any difference if you use something else?
   Alexey Zalesny: We tried also a weighted distance. In the beginning, we worked together with Georgyi Gimel'farb and he used another distance, but it was not critical. The results were stable.
2. **Bill Triggs, INRIA Rhône-Alpes:** Can you give us some intuition about the number of cliques that are needed to model typical natural textures?
   Alexey Zalesny: We tried to stop clique selection automatically. Sometimes we can do that during our stochastic probability synthesis. We can easily tell when to stop the generation. We can compare the biggest distance of the next not-selected clique type. If this is approximately the same as we already have, we can stop clique selection. We need about 40 cliques.
   Bill Triggs: So you can model almost any natural texture with about 40 cliques, at least at a single scale?
   Alexey Zalesny: Yes, even for colored texture it will be around 40 cliques distributed on 3 rasters. It means that for 32 gray level images we have 63 signal differences and 63x40 parameters. Typically, there are 1000, 2000, or 3000 parameters.
3. **Bill Triggs:** Secondly, how sensitive is the texture generation to the positions of the selected points? If you moved the points around locally and re-learned the statistics, would the generated textures change very much?
   Alexey Zalesny: The stability of this system is good. The sequential analysis of the algorithm is very good. When selecting the mutually dependent cliques we minimize the distance for all cliques. There is only a restriction on the maximal clique length. That is why if the points are moved slightly we get the same texture. We might find another neighborhood system, but this new system would give us the same result. We could still generate a similar texture.
4. **Andrew Fitzgibbon, University of Oxford:** You're using a greedy algorithm, selecting cliques sequentially. Would you expect to do better with a more global algorithm?
   Alexey Zalesny: At first, we tried to select simultaneously all cliques with the biggest histogram distance between the reference texture and independent random noise. However, the results of this synthesis were very bad. For this kind of analysis, we only need some milliseconds but the results were poor.
5. **Kyros Kutulakos, University of Rochester:** Can you say a little bit more about what you assume about the geometry of the surface over which you impose the texture when you try to change the viewpoint? Or if you wanted to texture map onto a curved surface and you wanted to change viewpoint, how would that change your warping function.
   Alexey Zalesny: For each additional view, we should know the geometry coefficients for a model of the translation invariance. Of course, we only introduce affine texture without perspective distortion. If you want to generate an orange, you should divide your orange into a finite number of oblique views, make a full analysis for one view and than quickly re-analyze for all the other

views. Now we need only the brightness information, not the height of the surface. The framework allows us to also use the height or other information. For example, for texture segmentation we just re-synthesize our map using the labels instead of the gray levels.

6. **Rudolf Mester, Frankfurt University:** You made a short reference to color texture synthesis and you mentioned that you were going to search for interactions between the RGB-planes. Because you want to keep the interactions as low as possible, wouldn't it be better to look for interactions in another color space like HSV or so?

Alexey Zalesny: I tried this but the results were not so good. We had some artifacts like unnatural spots of different colors. It was not so disturbing but nevertheless interesting to try. I tried in other color spaces like using two color difference signals and brightness but then the global colors were shifted.

Augmented Reality Using Uncalibrated Video Sequences

Kurt Cornelis*, Marc Pollefeys*, Maarten Vergauwen, and Luc Van Gool

K. U. Leuven, ESAT-PSI
Kardinaal Mercierlaan 94, B-3000 Leuven, Belgium
kcorneli | pollefey | vergauwe | vangool@esat.kuleuven.ac.be
WWW home page:http://www.esat.kuleuven.ac.be/~kcorneli

Abstract. *Augmented Reality*(AR) aims at merging the real and the virtual in order to enrich a real environment with virtual information. Augmentations range from simple text annotations accompanying real objects to virtual mimics of real-life objects inserted into a real environment. In the latter case the ultimate goal is to make it impossible to differentiate between real and virtual objects. Several problems need to be overcome before realizing this goal. Amongst them are the rigid registration of virtual objects into the real environment, the problem of mutual occlusion of real and virtual objects and the extraction of the illumination distribution of the real environment in order to render the virtual objects with this illumination model. This paper will unfold how we proceeded to implement an *Augmented Reality System* that registers virtual objects into a totally uncalibrated video sequence of a real environment that may contain some moving parts. The other problems of occlusion and illumination will not be discussed in this paper but are left as future research topics.

1 Introduction

1.1 Previous Work

Accurate registration of virtual objects into a real environment is an outspoken problem in Augmented Reality(AR). This problem needs to be solved regardless of the complexity of the virtual objects one wishes to enhance the real environment with. Both simple text annotations and complex virtual mimics of real-life objects need to be placed rigidly into the real environment. *Augmented Reality Systems* that lack this requirement will demonstrate serious 'jittering' of virtual objects in the real environment and will therefore fail to give the user a real-life impression of the augmented outcome.

The registration problem has already been tackled by several researchers in the AR-domain. A general discussion of all coordinate frames that need to be

---

* Kurt Cornelis and Marc Pollefeys are respectively research assistant and postdoctoral fellow of the Fund for Scientific Research - Flanders(Belgium)(F.W.O. - Vlaanderen)

registered with each other can be found in [25]. Some researchers use predefined geometric models of real objects in the environment to obtain vision-based object registration [15,22,27]. However, this delimits the application of such systems because geometric models of real objects in a general scene are not always readily available. Other techniques have been devised to make the calibration of the video camera obsolete by using affine object representations [16]. These techniques are simple and fast but fail to provide a real impression when projective skew is dominant in the video images. Therefore virtual objects can be viewed correctly only from large distances where the affine projection model is almost valid. So it seems that the most flexible registration solutions are those that don't depend on any a priori knowledge of the real environment and use the full perspective projection model. Our AR-System belongs to this class of flexible solutions.

To further enhance the real-life impression of an augmentation the occlusion and illumination problems need to be solved. The solutions to the occlusion problem are versatile. They differ in whether a 3D reconstruction of the real environment is needed or not [3,5]. Also the illumination problem has been handled in different ways. A first method uses an image of a reflective object at the place of insertion of the virtual object to get an idea of the incoming light at that point [6]. A second approach obtains the total reconstruction of a 3D radiance distribution by the same methods used to reconstruct a 3D scene [19]. Another approach consists of the approximation of the illumination distribution by a sphere of illumination directions at infinity [20].

As Computer Generated Graphics of virtual objects are mostly created with non physically-based rendering methods, techniques that use image-based rendering can be applied to incorporate real objects into another real environment [23] to obtain realistic results. Image-based rendering is explained in [7].

However, the 'jittering' of virtual objects in the real environment can degrade severely the final augmented result, even if problems of occlusion and illumination can be resolved exactly. We focussed on developing an AR-System that solves the registration problem as a prerequisite. It is based primarily on a 3D reconstruction scheme that extracts motion and structure from uncalibrated video images and uses the results to incorporate virtual objects into the real environment.

1.2 Overview

In the first upcoming section we will describe the motion and structure recovery algorithm of the AR-System. Although the main goal is the recovery of motion of the camera throughout the video sequence, the system also recovers a crude 3D structure of the real environment. This can be useful to handle future problems like resolving occlusions and extracting the illumination distribution of the real environment. We will focus on the motion recovery abilities of the AR-System.

In a following section we will discuss the use of the recovered motion parameters and the 3D structure to register virtual objects within the real environment.

This involves using the crude 3D representation of the real environment which we obtain as an extra from the motion recovery algorithm. Dense 3D reconstruction of the real environment is not necessary but may prove useful for future solutions to the occlusion problem.

Another section will give an overview of the final AR-algorithm. We will finish by showing results of the AR-System on some applications and by indicating future work to be done in order to upgrade the AR-System.

2 Motion and Structure Recovery

2.1 Preliminaries

As input to the AR-System we can take totally uncalibrated video sequences. The video sequences are neither preprocessed nor set up to contain calibration frames or fiducial markers in order to simplify motion and structure recovery. Extra knowledge on calibration parameters of the video camera can be used to help the AR-System to recover motion and structure but is not necessary to obtain good results.

The video sequences are not required to be taken from a purely static environment. As long as the moving parts in the real environment are small in the video sequence the algorithm will still be able to recover motion and structure.

2.2 Motion and Structure Recovery Algorithm

Image Features Selection and Matching Recovery of motion in Computer Vision is almost always based on tracking of features throughout images and uses these to determine motion parameters of the camera viewing the real environment. Features come in all flavours like points, lines, curves [4] or regions [26]. The features we use are the result of the Harris Corner Detector algorithm [9] applied to each image of our input video sequence. The result consists of points or *corners* in the images determining where the image intensity changes significantly in two orthogonal directions.

We end up with *corners* in each image of the video sequence but these are still unmatched from one image to another. We need to match them in different images in order to extract motion information. An initial set of possible matching corners is constructed using a small search region around each corner looking for corners in other images which have a large normalized intensity cross-correlation with the corner under scrutiny. Corresponding or *matching* corners are constrained through epipolar geometry to lie on each others epipolar line. This constraint can be expressed in terms of a linear equation between the two images one wishes to match the corners from:

$$x_1^T \mathbf{F}_{12} x_2 = 0 \qquad (1)$$

where $x_1 = (u_1, v_1, 1)^T$ and $x_2 = (u_2, v_2, 1)^T$ denote homogeneous image coordinates of matching corners in the first and second image. $\mathbf{F}_{12}$ is a 3×3

singular matrix which describes the epipolar geometry between the two images. The epipolar line from corner x_1 in image 2 and from corner x_2 in image 1 can be written down respectively as:

$$\mathbf{F}_{12}^T x_1 = 0 \text{ and} \quad (2)$$
$$\mathbf{F}_{12} x_2 = 0 \quad (3)$$

Using equation (1) each possible match between corners from the two images adds a constraint on the elements of the matrix $\mathbf{F}_{12}$. Extra constraints can be superimposed on $\mathbf{F}_{12}$ due to its singular nature and because it can only be determined up to a scalefactor as we are working with homogeneous image coordinates. Several algorithms have been devised to determine reliable matches between the corners of two images. These matches lead to a reasonable consistent $\mathbf{F}_{12}$, which means that equation (1) returns a small residual error for an important fraction of the presumed matches. The determination of this particular set of matches is achieved by a RANSAC algorithm [12] which determines $\mathbf{F}_{12}$ from trial matches and additional constraints of singularity and scalability. Once a good initial $\mathbf{F}_{12}$ is obtained it is optimized using all consistent matches and a Levenberg-Marquardt optimization technique.

As long as the moving parts in the real environment are small in the video sequence the RANSAC algorithm will treat corners belonging to these moving parts as outliers. They will be properly discarded in the determination of the matrix $\mathbf{F}_{12}$ and the matching corners.

Initializing Motion and Structure Recovery Once corner matches between two initial images are found, they can be used to initialize motion and structure recovery from the video sequence.

The relation between a 3D structure point and its projection onto an image can be described by a linear relationship in homogeneous coordinates:

$$m_k \sim \mathbf{P}_k M \quad (4)$$

in which $M = (X, Y, Z, 1)$ and $m_k = (x_k, y_k, 1)^T$ are the homogeneous coordinates of the 3D structure point and its projection onto image k respectively. $\mathbf{P}_k$ is a 3×4 matrix which describes the projection operation and '$\sim$' denotes that this equality is valid up to a scalefactor.

The two initial images of the sequence are used to determine a reference frame. The world frame is aligned with the camera of the first image. The second camera is chosen so that the epipolar geometry corresponds to the retrieved $\mathbf{F}_{12}$.

$$\begin{array}{l}\mathbf{P}_1 = [\quad \mathbf{I}_{3\times 3} \quad | \quad 0_3 \quad] \\ \mathbf{P}_2 = [\ [\mathbf{e}_{12}]_\times \mathbf{F}_{12} + \mathbf{e}_{12}\pi^\top \ | \ \sigma \mathbf{e}_{12}\ ]\end{array} \quad (5)$$

where $[\mathbf{e}_{12}]_\times$ indicates the vector product with $\mathbf{e}_{12}$. Equation (5) is not completely determined by the epipolar geometry (i.e. $\mathbf{F}_{12}$ and $\mathbf{e}_{12}$), but has 4 more degrees of freedom (i.e. π and σ). π determines the position of the reference plane (this corresponds to the plane at infinity in an affine or metric frame) and σ determines the global scale of the reconstruction. To avoid some problems during the reconstruction it is recommended to determine π in such a way that the reference plane does not cross the scene. Our implementation uses an approach similar to the quasi-Euclidean approach proposed in [2], but the focal length is chosen so that most of the points are reconstructed in front of the cameras[1]. This approach was inspired by Hartley's cheirality [10]. Since there is no way to determine the global scale from the images, σ can arbitrarily be chosen to $\sigma = 1$.

Once the cameras have been fully determined the matches can be reconstructed through triangulation. The optimal method for this is given in [11]. This gives us a preliminary reconstruction.

Updating Motion and Structure Recovery To obtain the matrix **P** or the corresponding motion of the camera for all other images in the video sequence a different strategy is used than the one described in the previous section.

First we take an image for which the corresponding matrix **P** has already been computed and retrieve the 2D-3D matches between corners in that image and the reconstructed 3D structure points. Secondly we take another image of which we only have the corners. With our RANSAC algorithm we compute the matrix **F** and corner matches between both images. Using corner matches between corners in image $k-1$ and image k and matches between corners in image $k-1$ and 3D structure points, we obtain matches between corners in image k and 3D structure points. See figure 1.

Knowing these 2D-3D matches we can apply a similar technique as we used to estimate **F**, to determine **P** taking into account equation (4) and a similar RANSAC algorithm. It is important to notice that the matrix **F** serves no longer to extract matrices **P**, but merely to identify corner matches between different images.

Using the previously reconstructed 3D structure points to determine **P** for the next image, we ensure that this matrix **P** is situated in the same projective frame as all previously reconstructed **P**'s. New 3D structure points can be initialized with the newly obtained matrix **P**. In this way the reconstructed 3D environment which one needs to compute **P** of the next image is updated on each step, enabling us to move all around a real object in a 3D environment if necessary.

In this manner motion and structure can be updated iteratively. However the next image to be calibrated cannot be chosen without care. Suppose one

---

[1] The quasi-Euclidean approach computes the plane at infinity based on an approximate calibration. Although this can be assumed for most intrinsic parameters, this is not the case for the focal length. Several values of the focal length are tried out and for each of them the algorithm computes the ratio of reconstructed points that are in front of the camera. If the computed plane at infinity −based on a wrong estimate of the focal length− passes through the object, then many points will end up behind the cameras. This procedure allows us to obtain a rough estimate of the focal length for the initial views.

Fig. 1. Knowing the corner matches between image k-1 and image k (m_{k-1}, m_k) and the 2D-3D matches for image k-1 (m_{k-1}, M), the 2D-3D matches for image k can be deduced (m_k, M).

chooses two images between which one wants to determine corner matches. If these images are 'too close' to each other, e.g. two consecutive images in a video sequence, the computation of the matrix **F** and therefore the determination of the corner matches between the two images becomes an ill-conditioned problem. Even if the matches could be found exactly the updating of motion and structure is ill-conditioned as the triangulation of newly reconstructed 3D points is very inaccurate as depicted in figure 2.

We resolved this problem by running through the video sequence a first time to build up an accurate but crude 3D reconstruction of the real environment. Accuracy is obtained by using keyframes which are separated sufficiently from each other in the video sequence. See figure 3. Structure and motion are extracted for these keyframes. In the next step each unprocessed image is calibrated using corner matches with the two keyframes between which it is positioned in the video sequence. For these new images no new 3D structure points are reconstructed as they will probably be ill-conditioned due to the closeness of the new image under scrutiny and its neighbouring keyframes. In this way a crude but accurate 3D structure is built up in a first pass along with the calibration of the keyframes. In a second pass, every other image is calibrated using the 2D-3D corner matches it has with its neighbouring keyframes. This leads to both a robust determination of the reconstructed 3D environment and the calibration of each image within the video sequence.

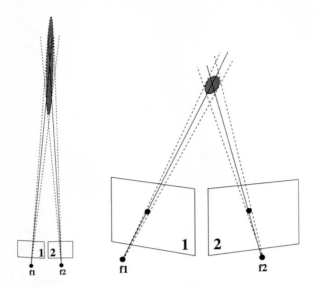

Fig. 2. left: If the images are chosen too close to each other the position and orientation of the camera hasn't changed much. Uncertainties in the image corners lead to a large uncertainty ellipsoid around the reconstructed point. Right: If images are taken further apart the camera position and orientation may differ more from one image to the next, leading to smaller uncertainty on the position of the reconstructed point.

Fig. 3. The small dots on the background represent the recovered crude 3D environment. The larger dark spots represent camera positions of keyframes in the video stream. The lighter spots represent the camera positions of the remaining frames.

Metric Structure and Motion Even for an uncalibrated camera some constraints on the intrinsic camera parameters are often available. For example, if the camera settings are not changed during recording, the intrinsic parameters will be constant over the sequence. In general, there is no skew on the image, the principal point is close to the center of the image and the aspect ratio is fixed (and often close to one). For a metric calibration the factorization of the **P**-matrices should yield intrinsic parameters which satisfy these constraints.

Self-calibration therefore consists of finding a transformation which allows the **P**-matrices to satisfy as much as possible these constraints. Most algorithms described in the literature are based on the concept of the absolute conic [8,24,18].

The presented approach uses the method described in [18]. The absolute conic ω is an imaginary conic located in the plane at infinity Π_∞. Both entities are the only geometric entities which are invariant under all Euclidean transformations. The plane at infinity and the absolute conic respectively encode the affine and metric properties of space. This means that when the position of Π_∞ is known in a projective framework, affine invariants can be measured. Since the absolute conic is invariant under Euclidean transformations its image only depends on the intrinsic camera parameters (focal length, ...) and not on the extrinsic camera parameters (camera pose). The following equation applies for the dual image of the absolute conic:

$$\omega_k^* \propto \mathbf{K}_k \mathbf{K}_k^\top \quad (6)$$

where $\mathbf{K}_k$ is an upper triangular matrix containing the camera intrinsics for image k. Equation (6) shows that constraints on the intrinsic camera parameters are readily translated to constraints on the dual image of the absolute conic. This image is obtained from the absolute conic through the following projection equation:

$$\omega_k^* \propto \mathbf{P}_k \Omega^* \mathbf{P}_k^\top \quad (7)$$

where Ω^* is the dual absolute quadric which encodes both the absolute conic and its supporting plane, the plane at infinity. The constraints on ω_k^* can therefore be back-projected through this equation. The result is a set of constraints on the position of the absolute conic (and the plane at infinity).

Our systems first uses a linear method to obtain an approximate calibration. This calibration is then refined through a non-linear optimization step in a second phase. More details on this approach can be found in [17].

3 Augmented Video

3.1 Virtual Object Embedding

Results obtained in the previous section can be used to merge virtual objects with the input video sequence. One can import the final calibration of each single image of the video sequence and the reconstructed crude 3D environment into a Computer Graphics System to generate augmented images.

In a Computer Graphics System virtual cameras can be instantiated which correspond to the retrieved calibrations of each image. The image calibrations include translation, rotation, focal length, principal point and skew of the actual real camera that took the image at that time. Typically Computer Graphics Systems do not support skew of the camera. This can easily be adapted in the software of the Computer Graphics System by including a skew transformation after performing the typical perspective transformation as explained in [13]. We use VTK [21] as our Computer Graphics Package. The virtual cameras can now be used to create images of virtual objects.

These virtual objects need to be properly registered with the real 3D environment. This is achieved in the following manner. First virtual objects are placed roughly within the 3D environment using its crude reconstruction. Fine-tuning of the position is achieved by viewing the result of a rough positioning by several virtual cameras and overlaying the rendering results from these virtual cameras on their corresponding real images in the video sequence. See figure 4. Using specific features in the real video images that were not reconstructed in the crude 3D environment a better and final placement of all virtual objects can be obtained. Note that at this stage of the implementation we don't take into account occlusions when rendering virtual objects.

3.2 Virtual Object Merging

After satisfactory placement of each single virtual object the virtual camera corresponding to each image is used to produce a virtual image. The virtual objects are rendered against a background that consists of the original real image. By doing so the virtual objects can be rendered with anti-aliasing techniques using the correct background for mixing.

4 Algorithm Overview

In this section the different steps taken by our AR-System are summarized :

step 1 : The initialization step. Take two images from the video sequence to initialize a projective frame in which both motion and structure will be reconstructed. During this initialization phase both images are registered within this frame and part of the 3D environment is reconstructed. One has to make sure these images are not taken too close or too far apart as this will lead to ill conditions. This is done by imposing a maximum and a minimum separation(counting number of frames) between the two images. The first image pair conforming to these bounds that leads to a good **F**-matrix is selected.

step 2 : Take the last image processed and another image further into the video sequence that still needs registering. Again these images are taken not too close or too far apart with the same heuristic method as applied in step 1.

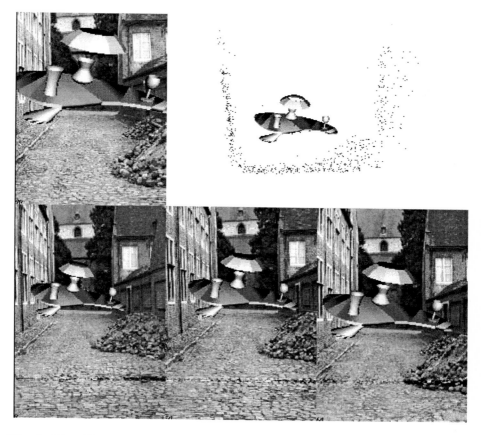

Fig. 4. The AR-interface : In the top right the virtual objects can be roughly placed within the crude reconstructed 3D environment. The result of this placement can be viewed instantaneously on some selected images.

step 3 : Corner matches between these images and the 2D-3D matches from the already processed image are used to construct 2D-3D matches for the image being registered.
step 4 : Using these new 2D-3D matches the matrix **P** for this image can be determined.
step 5 : Using **P** new 3D structure points can be reconstructed for later use.
step 6 : If the end of the video sequence is not reached, return to step 2.

Now only keyframes that are quite well separated have been processed. The remaining frames are processed in a manner similar to step 3 and 4.

step 7 : For each remaining frame the corner matches of the keyframes between which it lies and their 2D-3D matches are used to obtain 2D-3D matches for this frame.

step 8 : Similar to step 4, the matrix **P** of these frames can be calculated. However no additional 3D structure points are reconstructed.

Now all frames are registered and virtual objects can be placed into the real environment as described in section 3.

step 9 : First the virtual objects are roughly placed within the real environment using its crude 3D reconstruction obtained in previous steps.

step 10 : Finetuning of the positions of the virtual objects is done by seeing the result overlaid on some selected images and adjusting the virtual objects until satisfactory placement is obtained.

5 Examples

We filmed a sequence of a pillar standing in front of our department. Using the AR-System we placed a virtual box on top of this pillar. Note that by doing so we didn't have to solve the occlusion problem for now as the box was never occluded since we were looking down onto the pillar. The AR-System performed quite well. The 'jittering' of the virtual box on top of the pillar is still noticeable but very small. See figure 5.

Fig. 5. A virtual box is placed on top of a real pillar. 'Jittering' is still noticeable in the augmented video sequence but is very small.

Another example shows a walk through a street. The camera motion of the person taking the film was far from smooth. However the AR-System managed to register each camera position quite well. See figure 6.

Fig. 6. A street scene: The virtual box seems to stay firmly in place despite the jagged nature of the camera trajectory.

A third example shows another street scene but with a person walking around in it. Despite this moving real object the motion and structure recovery algorithm extracted the correct camera motion. See figure 7.

All video examples can be found at
http://www.esat.kuleuven.ac.be/~kcorneli/smile2.

6 Future Research

It is clear that the proposed AR-System can be further enhanced. One can try to reduce the 'jittering' of virtual objects by considering different techniques. E.g. incorporation of restrictions on the path followed by the real camera can be used to obtain a smoother path outlined by the virtual cameras. This leads to a smoother motion of the virtual objects in the augmented video and can therefore give more appealing results than the abrupt jumps in motion of noisy virtual camera positions. Another approach to reduce 'jittering' uses real image information in the neighbourhood of the virtual objects to lock it onto a real object. The latter technique is not useful in the case when virtual objects are meant to fly, float or move around in the real environment.

Fig. 7. Another street scene: Despite the moving person the motion of the camera can be extracted and used for augmenting the real environment with virtual objects.

The virtual objects used to augment the real environment can be the result of an earlier 3D reconstruction of real objects. A real vase could be modeled in a first 3D reconstruction step and the result used as virtual object to be placed on top of the real pillar. In this way expensive or fragile objects don't need to be handled physically to obtain the desired video. One can just use its 3D model instead and place it anywhere one wants in a real environment. E.g. relics or statues presently preserved in musea can be placed back in their original surrounding without endangering the precious original. This can be applied in producing documentaries or even a real-time AR-System at the archaeological site itself.

After the registration problem is solved in a satisfactory way we will dive into the occlusion and illumination problems which are still left to be solved and prove to be very challenging.

A topic which seems interesting is to simulate physical interactions between real and virtual objects. A simple form may be to implement a collision detection algorithm which can help us when placing virtual objects onto a surface of the real environment for easy positioning of the virtual objects.

7 Conclusion

In this paper we presented an AR-System which solves the registration problem of virtual objects into a video sequence of a real environment. It consists of two main parts.

viewpoint. We light a persons face from all different directions as quickly as possible using a spiraling light source on a 2-axis mechanism. In the course of a minute we get 2000 light directions. Suppose you want to illuminate the person with the illumination we extracted using our light-probe technique. You basically take these measurements of incident illumination, resample them to the granularity of the data set, modulate the two by each other and add all of the images together. So we just use image-based techniques to render the person under arbitrary illuminations like in Saint Peter's Basilica or in the forest. This is illustrated in Figure 3

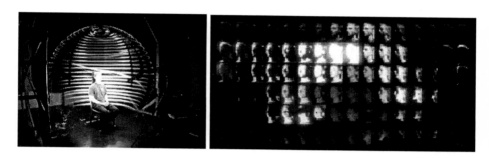

Fig. 3. Lightstage (left) and illumination modulated reflectance field (right).

That is a scene representation which allows you to illuminate a person quite efficiently. It doesn't require geometry of hair, or reflectance modeling of skin nor does it need global illumination algorithms. One question that arises when you wish to acquire a large scale environment with a technique like this is how to densely position light sources in 2D or 3D. I don't know about this yet.

Discussion

Jean Ponce, University of Illinois at Urbana-Champaign: I think there is a slight confusion. The lightfield is a geometric concept as well. I don't think you have to move away from geometry.

Paul Debevec: By geometry I mean surfaces, shapes. A lightfield doesn't assume shape. But it is a geometric concept because there are x,y,z-axes and things like that.

Kyros Kutulakos, University of Rochester (comment): I would like to give a comment on that. With a lightfield you cannot avoid the real problems, e.g. unless you have a good identification of shape, you need a huge amount of samples of the lightfield in order to create new photo-realistic pictures. So I'm not sure if you can clearly separate all these visual scene representations and choose one or the other. I think we're coming back to the same issue which is that while for years they are claiming that shape is not really required, in realistic situations we might have to start reasoning about shape again because it would help us with some practical issues.

Paul Debevec: That is exactly what saved us with 'Facade'. We were doing view-dependent texture mapping, which is a sparsely sampled lightfield. The only reason why it works with such sparsely sampled views is that we are assuming we have geometry. So that point is not lost. The question is if you are going to use the shape information in an explicit way. In a way that when you generate the renderings we can actually sense that that shape information is there. Or are you going to use the shape information to control the number of images you need to begin with or to do compression of the lightfield.

Andrew Fitzgibbon, University of Oxford: Let's look at what happens: you take say 100 images of the world to acquire some information. You then have a dataset and what you want to do is to exercise some degrees of freedom of that dataset. The one that we commonly exercise is camera position: we move the camera to different viewpoints. So the lightfield, which represents the world in terms of the camera viewpoints, allows you to index from that. So it is an ideal representation if your task is to change viewpoint.
But if you want to do something that interacts with the shape of the world, like putting an object on the table and then replaying the video, you do begin to require some representation of what the table was or what the depth was at that point in the images.

Marc Pollefeys: The first thing to do to tackle a problem is to acquire a huge amount of data and then you need some way to recombine the data to acquire the results you want. That's basically the lumigraph and plenoptic approach, and these are the simplest approaches in general. Then if you wanted to do more, you would have to recover the structure that is in the data. Like geometry, it is a structure that is present in the images. But the structure you find will never perfectly cover all the aspects of the phenomenon you want to describe. We will develop methods and insights to get to these structures which are intrinsically present within the data. This will allow us to do more things with the data than we can presently. We can extrapolate views when we have geometry.

Paul Debevec: Just to expand on that. Being able to extrapolate the plenoptic function is one thing, which is great because you then can fly around in the scene. Being able to delight the scene is another thing and we are not there yet. That means there will be many exciting results that we will be able to get as we become more able to do that with more and more complex scenes. But then we want to do more, like moving objects around in the scene. There is so much more to objects, like how much they weigh, how they feel when you hold them, what they smell like. As soon as we have systems that can display these things, we are becoming interested in these. So maybe we should start to think about a couple of things right now.

Hans-Helmut Nagel, Universität Karlsruhe: I wonder about the following question. The problem seems to be that you can completely make explicit many different properties in addition to the geometry and the reflection field for each point. But then the problem seems to be to determine the particular applications for which it may be more useful to keep some information implicit, and discover what does need to be explicit. The tradeoff is not yet clear to me. That is part of my question. Can we clearly state the conditions or tasks in which it is preferable that the data are implicit rather than making the data explicit and being able to interact with it?

The other point which came to me is that the topic of this discussion seems to be retrieval of the structure of the environment. I wonder to what extent the problem will shift again once you concentrate on people acting in that environment. Because then the attention of the viewer will be more on those people rather than on whether the environment is truly realistic. So it may be that the topic of this workshop will shift within the next decade. Studying the emphasis between acting people and how well they have to be represented versus how well we have to model the environment for a particular activity.

Sing Bing Kang: I would like to address your first question. I think that the representation should be a function of how complicated the object is. Say, for example, you want to model the interior of this theater, including this flat wall. I think it would be extremely wasteful if you used a purely image-based representation to do that when you can actually represent it by a simple planar surface, possibly with view-dependent textures. On the other hand, suppose you also have a very complicated object (such as a plant) whose geometry cannot be extracted accurately using stereo methods. It is probably best that you represent it using an image-based representation instead. As such, you can imagine an optimal rendering system being one that is capable of both model-based and image-based rendering, in order to take advantage of the merits of both representations.

References

1. P. Debevec, C. Taylor, and J. Malik. Modeling and rendering architecture from photographs: A hybrid geometry- and image-based approach. In *Proceedings of the ACM SIGGRAPH Conference on Computer Graphics*, pages 11–20, 1996.
2. S.B. Kang, R. Szeliski, and P. Anandan. The geometry-image representation tradeoff for rendering. In *International Conference on Image Processing, Vancouver, Canada*, page To appear, September 2000.
3. R. Koch, M. Pollefeys, B. Heigl, L. Van Gool, and H. Niemann. Calibration of handheld camera sequences for plenoptic modeling. In *Proc. International Conference on Computer Vision*, pages 585–591, 1999.
4. P. Rademacher. View-dependent geometry. In *Computer Graphics (SIGGRAPH'99), Los Angeles, CA*, pages 439–446, August 1999.

Geometry and Texture from Thousands of Images

J.P. Mellor*

Rose-Hulman Institute of Technology,
Terre Haute, IN 47803
j.p.mellor@rose-hulman.edu

Abstract. This paper presents a novel method for automatically recovering dense *surface patches* using large sets (1000's) of calibrated images taken from arbitrary positions within the scene. Physical instruments, such as Global Positioning System (GPS), inertial sensors, and inclinometers, are used to estimate the position and orientation of each image. Some of the most important characteristics of our approach are that it: 1) uses and refines noisy calibration estimates; 2) compensates for large variations in illumination; 3) tolerates significant soft occlusion (e.g. tree branches); and 4) associates, at a fundamental level, an estimated normal (eliminating the frontal-planar assumption) and texture with each surface patch.

1 Introduction

The problem of recovering three-dimensional information from a set of photographs or images is essentially the correspondence problem: *Given a point in one image, find the corresponding point in each of the other images.* Typically, photogrammetric approaches (Section 1.1) require manual identification of correspondences, while computer vision approaches (Section 1.2) rely on automatic identification of correspondences. If the images are from nearby positions and similar orientations (short baseline), they often vary only slightly, simplifying the identification of correspondences. Once sufficient correspondences have been identified, solving for the depth is simply a matter of geometry.

1.1 Photogrammetry

A number of the recent interactive modeling systems are based upon photogrammetry. Research projects such as RADIUS [2] and commercial systems such as FotoG are commonly used to extract three-dimensional models from images. Good results have been achieved with these systems, however the requirement

---

* This paper describes research done at the Artificial Intelligence Laboratory of the Massachusetts Institute of Technology and was supported in part by the Advanced Research Projects Agency of the Department of the Defense under Office of Naval Research contract N00014-91-J-4038 and Rome Lab contract F3060-94-C-0204.

for human input limits the size and complexity of the recovered model. One approach to reducing the amount of human input is to exploit geometric constraints. The geometric structure typical of urban environments can be used to constrain the modeling process such as Becker and Bove [1], Shum et al.[14], and Debevec et al. [7]. In spite of this reduction, each image must be processed individually by a human to produce a three-dimensional model, making it difficult to extend these systems to large sets of images. The major strength of these systems is the textured three-dimensional model produced.

1.2 Computer Vision

One approach to automatically solving the correspondence problem is to use multiple images such as Collins [3] and Seitz and Dyer [13] (neither of these approaches is suitable for reconstructions using images acquired from within the scene) or Kutulakos and Seitz [9,8] (which is not well suited to images acquired in outdoor urban environments).

1.3 Discussion

Photogrammetric methods produce high quality models, but require human input. Computer vision techniques function automatically, but generally do not produce usable models, operate on small sets of images and frequently are fragile with respect to occlusion and changes in illumination [1]. The work presented here draws from both photogrammetry and computer vision. Like photogrammetric methods we produce high quality textured models and like computer vision techniques our method is fully automatic.

Our approach is valid for arbitrary camera positions within the scene and is capable of analyzing very large sets of images. Our focus is recovering built geometry (architectural facades) in an urban environment. However, the algorithms presented are generally applicable to objects that can be modeled by small planar patches. Surface patches (geometry and texture) or *surfels* are recovered directly from the image data. In most cases, three-dimensional position and orientation can be recovered using purely local information, avoiding the computational costs of global constraints. Some of the significant characteristics of this approach are:

- Large sets of images contain both long and short baseline images and exhibit the benefits of both (accuracy and ease of matching). It also makes our method robust to sensor noise and occlusion, and provides the information content required to construct complex models.
- Each image is calibrated - its position and orientation in a single global coordinate system is estimated. The use of a global coordinate system allows data to be easily merged and facilitates geometric constraints. The camera's internal parameters are also known.

---

[1] For a more complete discussion of related work see [10].

– The fundamental unit is a textured surface patch and matching is done in three-dimensional space. This eliminates the need for the frontal-planar assumption made by many computer vision techniques and provides robustness to soft occlusion (e.g. tree branches). Also, the surface patches are immediately useful as a rough model and readily lend themselves to aggregation to form more refined models. The textured surface patch or *surface element* is referred to as a surfel. This definition differs from Szeliski's [15] in that it refers to a finite sized patch which includes both geometry and texture.
 – The algorithm tolerates significant noise in the calibration estimates and produces updates to those estimates.
 – The algorithm corrects for changes in illumination. This allows raw image properties (e.g. pixel values) to be used, avoiding the errors and ambiguities associated with higher level constructs such as edges or corners.
 – The algorithm scales well. The initial stage is completely local and scales linearly with the number of images. Subsequent stages are global in nature, exploit geometric constraints, and scale quadratically with the complexity of the underlying scene [2].

Not all of these characteristics are unique, but their combination produces a novel method of automatically recovering three-dimensional geometry and texture from large sets of images.

1.4 City Scanning Project

The work presented in this paper is part of the MIT City Scanning project whose primary focus is the Automatic Population of Geospatial Databases (APGD). A self contained image acquisition platform called *Argus* is used to acquire calibrated images [6]. At each location or node images are acquired in a hemispherical tiling. The position and orientation estimates obtained during the acquisition phase are good, but contain a significant amount of error. The estimates are refined using techniques described in [5,4]. For a more complete description of the project see [16,17].

1.5 Overview

Figure 1 shows an overview of the reconstruction pipeline described in this paper. The calibrated imagery described above serves as input to the pipeline. The left hand column shows the major steps of our approach; the right hand side shows example output at various stages. The output of the pipeline is a textured three-dimensional model. Our approach can be generally characterized as *hypothesize and test*. We hypothesize a surfel and then test whether it is consistent with the data. Section 2 describes the dataset used for this paper. Section 3 briefly reviews our approach. Section 4 introduces several techniques to remove false

---

[2] This is the worst case complexity. With spatial hashing the expected complexity is linear in the number of reconstructed surfels.

Geometry and Texture from Thousands of Images   173

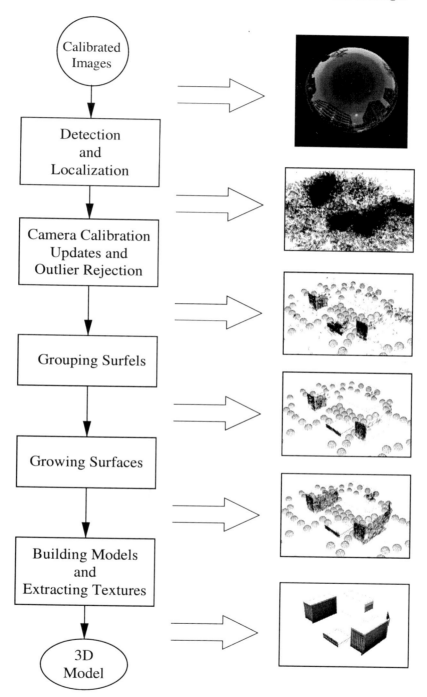

Fig. 1. Overview.

positives and fill in the missing parts. Finally we discuss fitting simple models and extracting textures. We present the results of applying our method to a large dataset.

2 The Dataset

A Kodak DCS 420 digital camera mounted on an instrumented platform was used to acquire a set of calibrated images in and around Technology Square [6]. Nearly 4000 images were collected from 81 node points. Other than avoiding inclement weather and darkness, no restrictions were placed on the day and time of, or weather conditions during, acquisition. The location of each node is shown in figure 2. At each node, the camera was rotated about the focal point collecting images in a hemispherical mosaic. Most nodes are tiled with 47 images. The raw images are 1524×1012 pixels and cover a field of view of $41° \times 28°$. Each node contains approximately 70 million pixels. After acquisition, the images are reduced to quarter resolution (381×253 pixels) and mosaiced [5,4]. Equal area projections of the spherical mosaic from two nodes are shown in Figure 3. The node on the left was acquired on an overcast day and has a distinct reddish tint. The one on the right was acquired on a bright clear day. Significant shadows are present in the right image whereas the left has fairly uniform lighting. Following mosaicing, the estimated camera calibrations are refined.

Fig. 2. Node locations.   **Fig. 3.** Example nodes.

After refinement, the calibration data is good, but not perfect. The pose estimates are within about 1° and 1 meter of the actual values. These errors produce an offset between corresponding points in different images. A 1° pose error will displace a feature by over 8 pixels. Our calibration estimates are in an absolute coordinate frame which allows us to integrate images regardless of when or from what source they were collected. This greatly increases the quantity and quality of available data, but because of variations in illumination condition also complicates the analysis.

Fig. 4. Reprojection onto surfel 1 coincident with actual surface.

Fig. 5. Source images for selected regions of surfel 1.

Figures 4 and 6 show several images from our dataset reprojected (using the estimated camera calibration) onto a surfel which is coincident with an actual surface. The location, orientation, and size of the surfels used are shown in Table 1. Surfel 1 was used to generate the collection of images in Figure 4 and surfel 2 those in Figure 6. If the camera calibration estimates were perfect and the illumination was constant, the regions in each figure should (ignoring errors introduced during image formation and resampling) be identical. The misalignment present in both sets is the result of error in the calibration estimates. Figure 4 is representative of the best in the dataset. A large number of the source images have high contrast and none of the regions are occluded. The third row has a distinct reddish tint. The four images in the center of the last row were collected under direct sunlight. And, the last two images were taken near dusk.

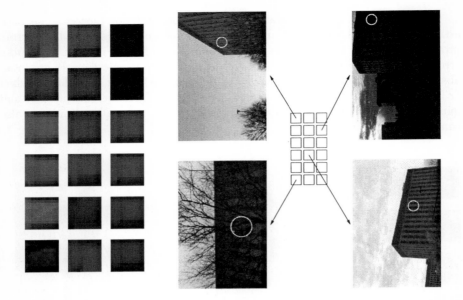

Fig. 6. Reprojection onto surfel 2 coincident with actual surface. **Fig. 7.** Source images for selected regions of surfel 2.

Figure 6 is more typical of the dataset. It is lower in contrast and some of the regions are partially occluded by trees. Figures 5 and 7 show source images with the reprojected area marked by a circle for several of the regions shown in Figures 4 and 6. In the worst cases all views of a surfel are similar to the upper left image of Figure 7.

3 Images to Surfels

Ideally, if a hypothesized surfel is coincident with an actual surface, reprojecting images onto the surfel should produce a set of regions which are highly correlated. On the other hand, if the surfel is not coincident with a surface, the subimages should not be correlated. This is the basic idea behind our approach [11]. As shown in Figures 4 and 6, noisy data is not quite this simple and we must extend our algorithm to handle camera calibration error, significant variation in illumination condition, and image noise (e.g. soft occlusion from tree branches) [12]. To compensate for camera calibration error we allow the source images to be shifted in the image plane prior to reprojection. We use optimization techniques to find the best alignment and the maximum shift is a function of the bound on calibration error. Illumination condition is normalized using a linear correction for each color channel. Finally, noisy pixels in the reprojection may individually be rejected as outliers. Figures 8 and 9 show the regions from Figures 4 and 6 after compensating for camera calibration error and illumination condition.

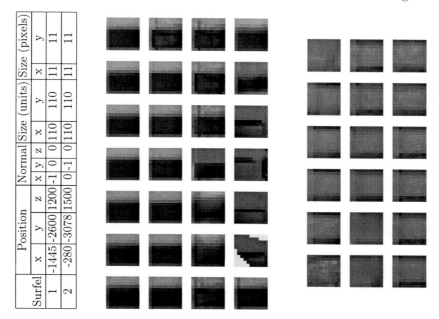

Surfel	Position			Normal			Size (units)		Size (pixels)	
	x	y	z	x	y	z	x	y	x	y
1	-1445	-2600	1200	-1	0	0	110	110	11	11
2	-280	-3078	1500	0	-1	0	110	110	11	11

Table 1. Surfel parameters. **Fig. 8.** Aligned and corrected regions for surfel 1. **Fig. 9.** Aligned and corrected regions for surfel 2.

Our extensions to handle camera calibration error, significant variation in illumination condition, and image noise add additional degrees of freedom to the problem making over-fitting a concern. Section 4 introduces several geometric constraints to prune false positives. One beneficial side effect is that surfels can be detected if they are simply near an actual surface. For hypothesized surfels which are within $\pm 30°$ and ± 100 units [3] of and actual surface, the detection rate is nearly 100%. We use the following algorithm to detect surfels.

1. Hypothesize a surfel in world coordinates.
2. Select images from cameras which can image the surfel.
3. Reproject the selected images onto the surfel.
4. Select one of the reprojected regions as a key region to match the others against.
5. For each region:
    a) Determine the shift which best aligns the region with the key region.
    b) Estimate color correction which produces the best match with the key region.
    c) Calculate best match with the key region.
6. Evaluate match set:
    – If good enough $\Rightarrow$ done.
    – If not $\Rightarrow$ goto 4.

Once a surface has been detected, the hypothesized position and orientation can be updated using the geometry of the matching regions and gradient information in the regions respectively. Detected surfels which are not false positives

---
[3] One unit is 0.1 foot.

typically converge to within 1 unit and a few degrees[4]. The detected surfel's position and orientation are localized using the following algorithm.

1. Until convergence:
   a) Update the surfel's position.
   b) Update surfel's orientation.
   c) Reevaluate the match set.

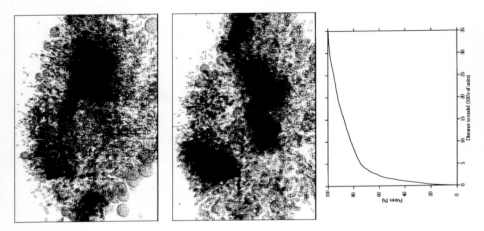

Fig. 10. Raw surfels.

Fig. 11. Distance to nearest model surface.

Figure 10 shows the results of applying the detection and localization algorithms to the dataset described in Section 2. Figure 11 shows the distribution of distances to the nearest model surfaces. Notice that while there are a significant number of false positives, most of the detected surfels are near actual surfaces.

4 Surfels to Surfaces

The results presented in the last section are purely local and make no attempt to reject false positives. This section explores several geometric constraints which together eliminate nearly all false positives.

4.1 Camera Updates

The shifts introduced in the last section are unconstrained. The actual image displacements caused by camera calibration error should be consistent with a translation of the camera center and a rotation about it. To enforce this constraint we use the following algorithm:

---

[4] For a more complete discussion of detection and localization see [10].

1. For each camera:
   a) Collect the shifts used to detect surfels.
   b) Calculate a first-order camera calibration update using these shifts and a linear least-squares solution.
   c) Use the first-order solution to filter the shifts.
   d) Calculate the final camera calibration update using non-linear optimization techniques.
   e) Remove matching regions with shifts that are not consistent with the final update.
2. Prune surfels which no longer meet the match criteria (i.e. too many matching regions have been removed because of inconsistent shifts).

This simple algorithm significantly improves the camera calibration (on average greater than a 3 fold improvement is achieved) and the remaining residuals are consistent with the nonlinear distortion which we have not modeled or corrected for. Figure 12 shows the consistent surfels remaining after applying this algorithm to the raw reconstruction shown in Figure 10 and Figure 13 shows the distribution of distances to the nearest model surface. A number of the consistent surfels come from objects which are not in the reference model. The cluster of surfels between the building outlines near the top center of Figure 12 is one example. These surfels come from a nearby building.

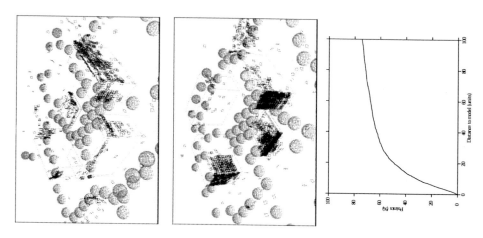

Fig. 12. Consistent surfels.

Fig. 13. Distribution of error for consistent surfels.

4.2 One Pixel One Surfel

Each pixel in each image should contribute to at most one surfel. Deciding which surfel is the hard part. Detection and localization as described in the last section do not enforce this constraint and as a result even after enforcing a consistent calibration update there are many image regions which contribute to multiple surfels. We eliminate them in a soft manner using the following algorithm.

1. Score each surfel based on the number of contributing cameras and the number of neighbors.
2. For each surfel S_a.
    a) For each region which contributes to S_a.
        i. For each surfel S_b with a score higher than S_a, if the region also contributes to S_b.
            A. De-weight the regions contribution to S_a.
    b) If the match score is no longer sufficient, prune S_a.

A surfel S_a is considered a neighbor of S_b if 1) the distance from the center of S_b to the plane containing S_a (the normal distance); 2) the distance from the projection of the center of S_b onto the plane containing S_a and the center of S_a (the tangential distance); and, 3) the angle between two orientations are all small enough. This notion of neighbors is essentially a smoothness constraint and is also used to group surfels. Figure 14 shows the reconstruction after pruning multiple contributions and Figure 15 shows the distribution of distances to the nearest model surface.

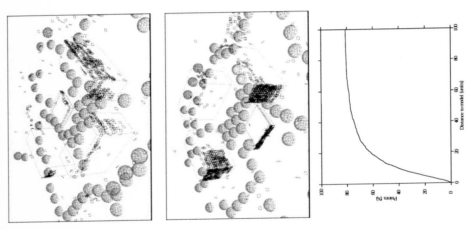

Fig. 14. Surfels after pruning multiple contributions.

Fig. 15. Distribution of error for pruned surfels.

4.3 Grouping Surfels

The buildings we are trying to model are much larger than an individual surfel. Therefore, a large number of surfels should be reconstructed for each actual surface. Using the notion of neighbors described above, we group the reconstructed surfels as follows:

1. For each surfel S_a.
    a) For each surfel S_b already assigned a group.

i. If S_a and S_b are neighbors.
      A. If S_a has not already been assigned to a group, then assign S_a to the group containing S_b.
      B. Otherwise merge the groups containing S_a and S_b.
   b) If S_a has not been assigned to a group, then create a new group and assign S_a to it.

In practice we retain only groups which have at least a minimum number of (typically four) surfels. All of the surfels in a group should come from the same surface. This notion of grouping places no restriction on the underlying surface other than smoothness (e.g. it may contain compound curves). Figure 16 shows the reconstruction after grouping and removing groups with fewer than four surfels. Nearly all of the surfaces in the reference model have at least one corresponding group. Figure 17 shows the distribution of distances to the nearest model surface.

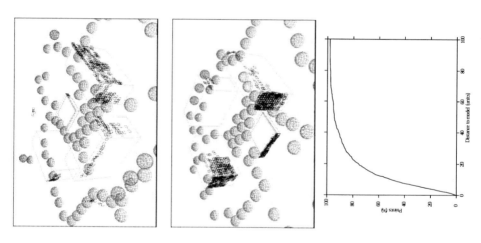

Fig. 16. Surfels after grouping.

Fig. 17. Distribution of error for grouped surfels.

4.4 Growing Surfaces

Many of the groups shown in figure 16 do not completely cover the underlying surface. There are several reasons why surfels corresponding to actual surfaces might not produce a valid match set. The main one is soft occlusion from tree branches. Another is local maxima encountered while finding the best shifts and updating the surfel's normal. We use the following algorithm to grow surfaces:

1. For each group.
   a) Create an empty list of hypothesized surfels.
   b) For each hypothesized surfel.

        i. Test using the detection and localization algorithms.
       ii. If a match.
           A. Add to the group.
           B. Test against each surfel in each of the other groups.
              If a neighbor, merge the two groups.
    c) Use the next surfel in the group S_a to generate new hypotheses
       and goto Step 1b.

The hypotheses in Step 1c are generated from S_a by considering the eight nearest neighbors in the plane containing S_a. The shifts and illumination corrections associated with S_a are used as initial values for each hypothesis in Step 1(b)i. Figure 18 shows the reconstruction after growing. After growing, the coverage of each surface is nearly complete. Figure 19 shows the distribution of distances to the nearest model surface.

4.5 Extracting Models and Textures

So far, the only assumption we have made about the structure of the world is that locally it can be approximated by a plane. All of the buildings imaged in our dataset are composed of planar faces, therefore we simply fit planes to the groups identified in the previous section. In this case, a face is equivalent to a large surfel. Figure 20 shows the reconstructed faces. A total of 15 surfaces were recovered. Figure 21 shows the distribution of distances to the nearest model surface. As noted previously, many surfels come from structures not in the reference model. Three of the reconstructed surfaces fall into this category, hence Figure 21 has a maximum of 80.

Using the illumination corrections calculated during detection and localization we can transform the images which contribute to a face into a common color space. To obtain the texture associated with each face, we simply reproject the color corrected images and average the results. Figure 22 shows two views of the reconstructed textures. Notice that the rows of window in adjacent faces are properly aligned. This occurs even though no constraints between faces are imposed.

4.6 Discussion

This section uses several simple geometric constraints to remove virtually all false positives from the purely local reconstruction described in Section 3. After imposing consistent calibration updates, removing multiple contributions and grouping, the remaining surfels are excellent seeds for growing surfaces. Of the 16 surfaces in the reference model, 12 were recovered. All of the remaining surfaces are severely occluded by trees. Nearly all of the images are similar to the upper left-hand image of Figures 6 and 7. In spite of this several surfels were recovered on two of these surfaces, however they did not survive the grouping process. In addition to being severely occluded by trees, the other two surfaces have very little texture and one of them suffers from a lack of data. Three surfaces from

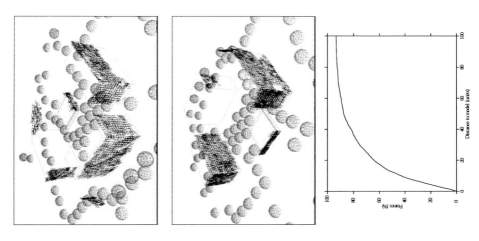

Fig. 18. Surfels after growing.

Fig. 19. Distribution of error for grown surfels.

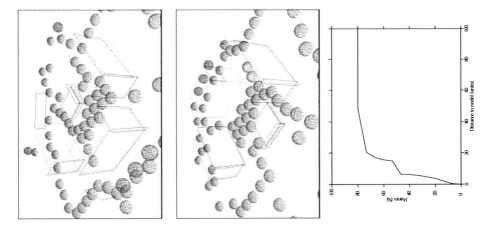

Fig. 20. Raw model surfaces.

Fig. 21. Distribution of error for model surfaces.

Fig. 22. Textured model surfaces.

adjacent buildings not contained in the model were also recovered. The face near the top center of the upper image in Figure 22 is from the Parson's lab. The surfaces on the left of the upper and the right of the lower image is from Draper lab.

5 Conclusions

This paper presents a novel method for automatically recovering dense surfels using large sets (1000's) of calibrated images taken from arbitrary positions within the scene. Physical instruments, such as Global Positioning System (GPS), inertial sensors, and inclinometers, are used to estimate the position and orientation of each image. Long baseline images improve the accuracy; short baselines and the large number of images simplify the correspondence problem. The initial stage of the algorithm is completely local enabling parallelization and scales linearly with the number of images. Subsequent stages are global in nature, exploit geometric constraints, and scale quadratically with the complexity of the underlying scene [5].

We describe techniques for:

- Detecting and localizing surfels.
- Refining camera calibration estimates and rejecting false positive surfels.
- Grouping surfels into surfaces.
- Growing surfaces along a two-dimensional manifold.
- Producing high quality, textured three-dimensional models from surfaces.

Some of our approach's most important characteristics are:

- It is fully automatic.
- It uses and refines noisy calibration estimates.
- It compensates for large variations in illumination.
- It matches image data directly in three-dimensional space.
- It tolerates significant soft occlusion (e.g. tree branches).
- It associates, at a fundamental level, an estimated normal (eliminating the frontal-planar assumption) and texture with each surfel.

Our algorithms also exploit several geometric constraints inherent in three-dimensional environments and scales well to large sets of images. We believe that these characteristics will be important for systems which automatically recover large-scale high-quality three-dimensional models. A set of about 4000 calibrated images was used to test our algorithms. The results presented demonstrate that they can be used for three-dimensional reconstruction. To our knowledge, the City Scanning project (e.g. [4] and the work presented in this paper) is the first to produce high-quality textured models from such large image sets. The image sets used are nearly two orders of magnitude larger than the largest sets used by other approaches. The approach presented in this paper, recovering dense surfels by matching raw image data directly in three-dimensional space, is unique among the City Scanning approaches.

---

[5] For a complete discussion of complexity and future work see [10].

References

1. Shawn Becker and V. Michael Bove, Jr. Semiautomatic 3-D model extraction from uncalibrated 2-D camera views. In *Visual Data Exploration and Analysis*, volume 2410, pages 447–461. SPIE, February 1995. San Jose, CA.
2. R. Collins, C. Jaynes, F. Stolle, X. Wang, Y. Cheng, A. Hanson, and E. Riseman. A system for automated site model acquisition. In *Integrating Photogrammetric Techniques with Scene Analysis and Machine Vision II*, volume 7617. SPIE, April 1995. Orlando, FL.
3. Robert T. Collins. A space-sweep approach to true multi-image matching. In *Computer Vision and Pattern Recognition (CVPR '96 Proceedings)*, pages 358–363, June 1996. San Francisco, CA.
4. Satyan Coorg. *Pose Imagery and Automated 3-D Modeling of Urban Environments.* PhD thesis, MIT, 1998.
5. Satyan Coorg, Neel Master, and Seth Teller. Acquisition of a large pose-mosaic dataset. In *Computer Vision and Pattern Recognition (CVPR '98 Proceedings)*, pages 872–878, June 1998. Santa Barbara, CA.
6. Douglas S. J. De Couto. Instrumentation for rapidly acquiring pose-imagery. Master's thesis, MIT, 1998.
7. Paul E. Debevec, Camillo J. Taylor, and Jitendra Malik. Modeling and rendering architecture from photographs: A hybrid geometry- and image-based approach. In *Computer Graphics (SIGGRAPH '96 Proceedings)*, pages 11–20, August 1996.
8. Kiriakos N. Kutulakos and Steven M. Seitz. A theory of shape by space carving. CS Technical Report 692, University of Rochester, May 1998.
9. Kiriakos N. Kutulakos and Steven M. Seitz. What do n photographs tell us about 3D shape? CS Technical Report 680, University of Rochester, January 1998.
10. J.P. Mellor. Automatically recovering geometry and texture from large sets of calibrated images. Technical Report AITR–1674, MIT, October 1999.
11. J.P. Mellor, Seth Teller, and Tomás Lozano-Pérez. Dense depth maps from epipolar images. In *Image Understanding Workshop (IUW '97 Proceedings)*, volume 2, pages 893–900, May 1997. New Orleans, LA.
12. J.P. Mellor, Seth Teller, and Tomás Lozano-Pérez. Dense surface patches from thousands of pose images. In *Image Understanding Workshop (IUW '98 Proceedings)*, volume 2, pages 537–542, November 1998. Monterey, CA.
13. Steven M. Seitz and Charles R. Dyer. Photorealistic scene reconstruction by voxel coloring. In *Computer Vision and Pattern Recognition (CVPR '97 Proceedings)*, pages 1067–1073, 1997. Puerto Rico.
14. Heung-Yeung Shum, Mei Han, and Richard Szeliski. Interactive construction of 3D models from panoramic mosaics. In *Computer Vision and Pattern Recognition (CVPR '98 Proceedings)*, pages 427–433, June 1998. Santa Barbara, CA.
15. Richard Szeliski and David Tonnesen. Surface modeling with oriented particle systems. In *Computer Graphics (SIGGRAPH '92 Proceedings)*, volume 26, pages 185–194, July 1992.
16. Seth Teller. Automated urban model acquisition: Project rationale and status. In *Image Understanding Workshop (IUW '98 Proceedings)*, volume 2, pages 455–462, November 1998. Monterey, CA.
17. Seth Teller. Toward urban model acquisition from geo-located images. In *Pacific Graphics '98*, 1998. Singapore.

Discussion

1. **Stephan Heuel, Bonn University**: I have a question about the grouping stage and the rejection stage: If you consider not only orthogonal walls but walls which have different angles, like 40 or 30 degrees, what happens to the rejection and the grouping ?

   J. P. Mellor: Outliers are pruned at several stages. The angle between two walls has no direct effect on the pruning. For example, most of the outliers are rejected during the camera calibration update. This is accomplished by imposing a consistent camera calibration and makes no assumptions about the structure of the world. Grouping, on the other hand, is essentially a smoothness constraint. We consider surfels which are spatially close to each other and have orientations within about 20 degrees to come from the same surface. The only other constraint imposed by grouping is that valid surfaces must contain at least four surfels. The smoothness constraint of the grouping stage is actually imposed during model fitting. In this stage we simply fit planes to the groups. Clearly, if two walls are within about 20 degrees this simplistic modeling will fail. More sophisticated modeling may help and I should point out that the raw surfels (after grouping) could be used as a rough model.

2. **Andrew Davison, University of Oxford**: How do you actually determine the orientation of your patches?

   J. P. Mellor: We took the brute force approach—we simply voxelize the area of interest and test them all. Our surfel detection technique can detect and localize surfaces that are within about 100 units (10 feet) and 30 degrees of the test point. So we sample every 100 units and 45 degrees. There certainly are smarter ways of generating test points and this is an area we are exploring.

VideoPlus: A Method for Capturing the Structure and Appearance of Immersive Environments

Camillo J. Taylor

GRASP Laboratory, CIS Department
University of Pennsylvania
3401 Walnut Street, Rm 335C
Philadelphia, PA, 19104-6229
cjtaylor@central.cis.upenn.edu
Phone: (215) 898 0376
Fax: (215) 573 2048

Abstract. This paper describes an approach to capturing the appearance and structure of immersive environments based on the video imagery obtained with an omnidirectional camera system. The scheme proceeds by recovering the 3D positions of a set of point and line features in the world from image correspondences in a small set of key frames in the image sequence. Once the locations of these features have been recovered the position of the camera during every frame in the sequence can be determined by using these recovered features as fiducials and estimating camera pose based on the location of corresponding image features in each frame. The end result of the procedure is an omnidirectional video sequence where every frame is augmented with its pose with respect to an absolute reference frame and a 3D model of the environment composed of point and line features in the scene.
By augmenting the video clip with pose information we provide the viewer with the ability to navigate the image sequence in new and interesting ways. More specifically the user can use the pose information to travel through the video sequence with a trajectory different from the one taken by the original camera operator. This freedom presents the end user with an opportunity to immerse themselves within a remote environment.

1 Introduction

This paper describes an approach to capturing the appearance and structure of immersive environments based on the video imagery obtained with an omnidirectional camera system such as the one proposed by Nayar [15]. The scheme proceeds by recovering the 3D positions of a set of point and line features in the world from image correspondences in a small set of key frames in the image sequence. Once the locations of these features have been recovered the position of the camera during every frame in the sequence can be determined by using these recovered features as fiducials and estimating camera pose based on the location of corresponding image features in each frame. The end result of the procedure is an omnidirectional video sequence where every frame is augmented with its pose with respect to an absolute reference frame and a 3D model of the environment composed of point and line features in the scene.

One area of application for the proposed reconstruction techniques is in the field of virtual tourism. By augmenting the video clip with pose information we provide the viewer with the ability to navigate the image sequence in new and interesting ways. More specifically the user can use the pose information to travel through the video sequence with a trajectory different from the one taken by the original camera operator. This freedom presents the end user with an opportunity to immerse themselves within a remote environment and to control what they see.

Another interesting application of the proposed technique is in the field of robotics since it allows us to construct 3D models of remote environments based on the video imagery acquired by a mobile robot. For example, the model of an indoor environment shown in Figure 5 was constructed from the video imagery acquired by the mobile robot shown in Figure 14c as it roamed through the scene.

Such a model would allow the remote operator to visualize the robots operating environment. It could also be used as the basis for an advanced human robot interface where the robot could be tasked by pointing to a location on the map and instructing it to move to that position. The robot would be able to automatically plan and execute a collision free path to the destination based on the information contained in the map.

The rest of this paper is arranged as follows Section 2 describes the process whereby the 3D locations of the model features and the locations of the cameras are estimated from image measurements. Results obtained by applying these techniques to actual video sequences are presented in Section 3. Section 4 discusses the relationship between this research and previously published work. Section 5 briefly describes future directions of this research and section 6 presents some of the conclusions that have been drawn so far.

2 Reconstruction

This section describes how the 3D structure of the scene and the locations of the camera positions are recovered from image correspondences in the video sequence. The basic approach is similar in spirit to the reconstruction schemes described in [20] and [4]. The reconstruction problem is posed as an optimization problem where the goal is to minimize an objective function which indicates the discrepancy between the predicted image features and the observed image features as a function of the model parameters and the camera locations.

In order to carry out this procedure it is important to understand the relationship between the locations of features in the world and the coordinates of the corresponding image features in the omnidirectional imagery. The catadioptric camera system proposed by Nayar [15] consists of a parabolic mirror imaged by an orthographic lens. With this imaging model there is an effective single point of projection located at the focus of the parabola as shown in Figure 1.

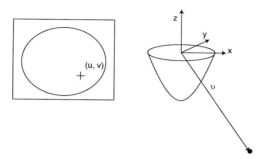

Fig. 1. The relationship between a point feature in the omnidirectional image and the ray between the center of projection and the imaged point.

Given a point with coordinates (u, v) in the omnidirectional image we can construct a vector, v, which is aligned with the ray connecting the imaged point and the center of projection of the camera system.

$$v = \begin{pmatrix} s_x(\mathrm{u} - c_x) \\ s_y(\mathrm{v} - c_y) \\ (s_x(\mathrm{u} - c_x))^2 + (s_y(\mathrm{v} - c_y))^2 - 1 \end{pmatrix} \quad (1)$$

This vector is expressed in terms of a coordinate frame of reference with its origin at the center of projection and with the z-axis aligned with the optical axis of the device as shown in Figure 1.

The calibration parameters, s_x, s_y, c_x and c_y associated with the imagery can be obtained in a separate calibration procedure [5]. It is assumed that these calibration parameters remain constant throughout the video sequence.

Note that since the catadioptric camera system has a single point of projection it is possible to resample the resulting imagery to produce "normal" perspective with arbitrary viewing directions [15]. The current system exploits this capability by providing a mechanism which allows the user to create a virtual viewpoint which she can pan and tilt interactively.

The current implementation of the reconstruction system allows the user to model two types of features: point features and straight lines aligned with one of the vertical or horizontal axes of the global frame of reference. These types of features were chosen because they are particularly prevalent and salient in man-made immersive environments but other types of features, such as lines at arbitrary orientations, could easily be included. The locations of point features can be represented in the usual manner by three vectors (X_i, Y_i, Z_i) [1] while the locations of the straight lines can be denoted with only two parameters. For example, the location of a vertical line can be specified by parameterizing the location of its intercept with the xy-plane (X_i, Y_i) since the vertical axis corresponds to the z-axis of the global coordinate frame. Note that for the purposes of reconstruction the lines are considered to have infinite length so no attempt is made to represent their endpoints.

The position and orientation of the camera with respect to the world frame of reference during frame j of the sequence is captured by two parameters, a rotation $R_j \in SO(3)$ and a translation $\mathbf{T}_j \in \mathbb{R}^3$. This means that given the coordinates of a point in the global coordinate frame, $\mathbf{P}_{iw} \in \mathbb{R}^3$ we can compute its coordinates with respect to camera frame j, $\mathbf{P}_{ij}$ from the following expression.

$$\mathbf{P}_{ij} = R_j(\mathbf{P}_{iw} - \mathbf{T}_j) \qquad (2)$$

The reconstruction program takes as input a set of correspondences between features in the omnidirectional imagery and features in the model. For correspondences between point features in the image and point features in the model we can construct an expression which measures the discrepancy between the predicted projection of the point and the vector obtained from the image measurement, v_{ij}, where $\mathbf{P}_{ij}$ is computed from equation 2.

$$\|(v_{ij} \times \mathbf{P}_{ij})\|^2 / (\|\mathbf{P}_{ij}\|^2 \|v_{ij}\|^2) \qquad (3)$$

This expression yields a result equivalent to the square of the sine of the angle between the two vectors, v_{ij} and $\mathbf{P}_{ij}$ shown in Figure 2.

For correspondences between point features in the image and line features in the model we consider the plane containing the line and the center of projection of the image. The normal to this plane, $\mathbf{m}_{ij}$ can be computed from the following expression.

$$\mathbf{m}_{ij} = R_j(\mathbf{v}_i \times (\mathbf{d}_i - \mathbf{T}_j)) \qquad (4)$$

Where the vector $\mathbf{v}_i$ denotes the direction of the line in space and the vector $\mathbf{d}_i$ denotes an arbitrary point on the line. As an example, for vertical lines the vector $\mathbf{v}_i$ will be aligned with the z axis $(0, 0, 1)^T$ and the vector $\mathbf{d}_i$ will have the form $(X_i, Y_i, 0)^T$.

The following expression measures the extent to which the vector obtained from the point feature in the omnidirectional imagery, v_{ij}, deviates from the plane defined by the vector $\mathbf{m}_{ij}$.

$$(\mathbf{m}^T v_{ij})^2 / (\|\mathbf{m}_{ij}\|^2 \|v_{ij}\|^2) \qquad (5)$$

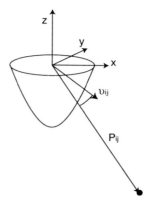

Fig. 2. Given a correspondence between a point feature in the omnidirectional image and a point feature in the model we can construct an objective function by considering the disparity between the predicted ray between the camera center and the point feature, $\mathbf{P}_{ij}$, and the vector v_{ij} computed from the image measurement.

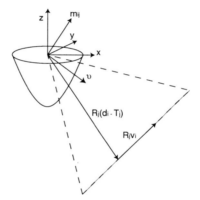

Fig. 3. Given a correspondence between a point feature in the omnidirectional image and a line feature in the model we can construct an objective function by considering the disparity between the predicted normal vector to the plane containing the center of projection and the model line, $\mathbf{m}_{ij}$, and the vector, v_{ij}, computed from the image measurement.

A global objective function is constructed by considering all of the correspondences in the data set and summing the resulting expressions together. Estimates for the structure of the scene and the locations of the cameras are obtained by minimizing this objective function with respect to the unknown parameters, R_j, $\mathbf{T}_j$, X_i, Y_i and Z_i. This minimization is carried out using a variant of the Newton-Raphson method [19,20,10].

---

[1] the subscript i serves to remind us that these parameters describe the position of the ith feature in the model.

An initial estimate for the orientation of the camera frames, R_j, can be obtained by considering the lines in the scene with known orientation such as lines parallel to the x, y, or z axes of the environment. If v_1 and v_2 represent the vectors corresponding to two points along the projection of a line in the image plane then the normal to the plane between them in the cameras frame of reference can be computed as follows $\mathbf{n} = v_1 \times v_2$. If R_j represents the rotation of the camera frame and $\mathbf{v}$ represents the direction of the line in world coordinates then the following objective function represents the fact that the normal to the plane should be perpendicular to the direction of the line in the coordinates of the camera frame.

$$(\mathbf{n}^T R_j \mathbf{v})^2 \tag{6}$$

An objective function can be created by considering all such lines in an image and summing these penalty terms. The obvious advantage of this expression is that the only unknown parameter is the camera rotation R_j which means that we can minimize the expression with respect to this parameter in isolation to obtain an initial estimate for the camera orientation.

The current implementation of the reconstruction system also allows the user to specify constraints that relate the features in the model. For example the user would be able to specify that two or more features share the same z-coordinate which would force them to lie on the same horizontal plane. This constraint is maintained by reparameterizing the reconstruction problem such that the z-coordinates of the points in question all refer to the same variable in the parameter vector.

The ability to specify these relationships is particularly useful in indoor environments since it allows the user to exploit common constraints among features such as two features belonging to the same wall or multiple features lying on a ground plane. These constraints reduce the number of free parameters that the system must recover and improve the coherence of the model when the camera moves large distances in the world.

Using the procedure outlined above we were able to reconstruct the model shown in Figure 5 from 14 images taken from a video sequence of an indoor scene.

The polyhedral model is constructed by manually attaching surfaces to the reconstructed features. Texture maps for these surfaces are obtained by sampling the original imagery.

An important practical advantage of using omnidirectional imagery in this application is that the 3D structure can be recovered from a smaller number of images since the features of interest are more likely to remain in view as the camera moves from one location to another.

Once the locations of a set of model features have been reconstructed using the image measurements obtained from a set of keyframes in the sequence, these features can then be used as fiducials to recover the pose of the camera at other frames in the sequence.

VideoPlus: A Method for Capturing Structure and Appearance   193

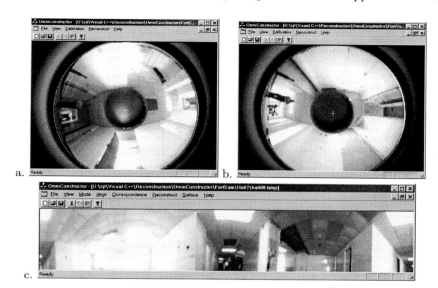

Fig. 4. Two of the omnidirectional images from a set of 14 keyframes are shown in a and b. A panoramic version of another keyframe is shown in c.

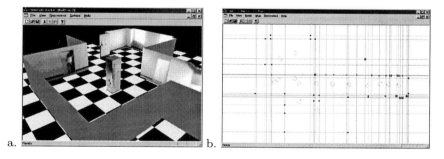

Fig. 5. a. 3D model of the environment constructed from the data set shown in Figure 4. b. Floor plan view showing the estimated location of all the images and an overhead view of the feature locations. The circles correspond to the recovered camera positions while the dots and crosses correspond to vertical line and point features.

For example, if frame number 1000 and frame number 1500 were used as keyframes in the reconstruction process then we know where a subset of the model features appears in these frames. Correspondences between features in the intervening images and features in the model can be obtained by applying applying standard feature tracking algorithms to the data set. The current system employs a variant of the Lucas and Kanade [14] algorithm to localize and track feature points through intervening frames.

Based on these correspondences, the pose of the camera during these intermediate frames can be estimated by simply minimizing the objective function described previously with respect to the pose parameters of the camera. The locations of the feature points are held constant during this pose estimation step. Initial estimates for the camera pose can be obtained from the estimates for the locations of the keyframes that were produced during the reconstruction process.

Another approach to estimating the pose of the camera during the intervening frames is to simply interpolate the pose parameters through the frames of the subsequence. That is, given that the camera pose in frames 1000 and 1500 is known we could simply estimate the roll, pitch and yaw angles of the intervening frames along with the translational position by interpolating these parameter values linearly. This approach is most appropriate in situations where the camera is moving with an approximately constant translational and angular velocity between keyframes.

Once the video sequence has been fully annotated with camera pose information the user is able index the data set *spatially* as well as temporally. In the current implementation the user is able to navigate through an immersive environment such as the office complex shown in Figure 6 in a natural manner by panning and tilting his virtual viewpoint and moving forward and backward. As the user changes the location of her viewpoint the system simply selects the closest view in the omnidirectional video sequence and generates an image in the approriate viewing direction.

The system also allows the user to generate movies by specifying a sequence of keyframes. The system automatically creates the sequence of images that correspond to a smooth camera trajectory passing through the specified positions. This provides the user with the capability of reshooting the scene with a camera trajectory which differs from the one that was used to capture the video initially.

3 Results

In order to illustrate what can be achieved with the proposed techniques we present results obtained from three different immersive environments.

Fig. 6. Three images taken from a video sequence obtained as the camera is moved through the library.

Figure 6 shows three images taken from a video sequence acquired in the Fine Arts Library at the University of Pennsylvania. This building was designed

Fig. 7. Images of the Fine Arts library at the University of Pennsylvania. The building was designed by Frank Furness in 1891 and remains one of the most distinctive and most photographed buildings on campus.

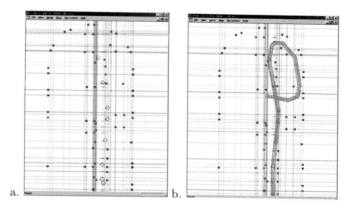

Fig. 8. a. A floor plan view of the library showing the locations of the features recovered from 9 keyframes in the video sequence. The circles correspond to the recovered camera positions while the dots and crosses correspond to line and point features. b. Based on these fiducials the system is able to estimate the location of the camera for all the intervening frames.

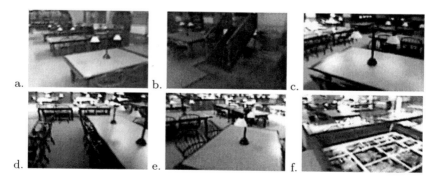

Fig. 9. Views generated by the system as the user conducts a virtual tour of the library.

by Frank Furness in 1891 and refurbished on its centenary in 1991, images of the interior and exterior of the building are shown in Figure 7.

The reconstruction of this environment was carried out using approximately 100 model features viewed in 9 frames of the video sequence. Figure 8a shows a floor plan view of the resulting reconstruction. The reconstructed feature locations were then used as fiducials to recover the position of 15 other frames in the sequence. Pose interpolation was employed to estimate the position and orientation of the camera during intervening frames. Figure 8b shows the resulting estimates for the camera position during the entire sequence. The original video sequence was 55 seconds long and consisted of 550 frames. During the sequence the camera traveled a distance of approximately 150 feet. Figure 9 shows viewpoints generated by the system as the user conducts a virtual tour of this environment.

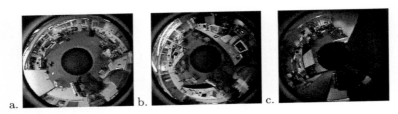

Fig. 10. Three images taken from a video sequence obtained as the camera is moved through the GRASP laboratory.

Fig. 11. Images of the GRASP laboratory at the University of Pennsylvania.

Figure 10 shows three images taken from a video sequence acquired in the GRASP laboratory at the University of Pennsylvania; snapshots of the lab are shown in Figure 11. In this case the video imagery was obtained in a sequence of short segments as the camera was moved through various sections of the laboratory. The entire video sequence was 154 seconds long and consisted of 4646 frames. The approximate dimensions of the region of the laboratory explored are 36 feet by 56 feet and the camera moved over 250 feet during the exploration. The reconstruction of this scene was carried out using approximately 50 model features viewed in 16 images of the sequence. The resulting model is shown

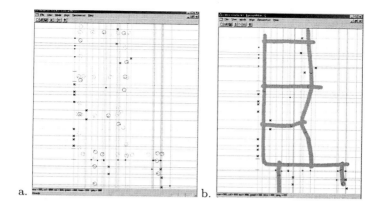

Fig. 12. a. A floor plan view of the laboratory showing the locations of the features recovered from 17 keyframes in the video sequence. The circles correspond to the recovered camera positions while the dots and crosses correspond to line and point features. b. Based on these fiducials the system is able to estimate the location of the camera for all the intervening frames. Notice that during the exploration the camera is moved into two side rooms that are accessed from the corridor surrounding the laboratory; these are represented by the two excursions at the bottom of this figure.

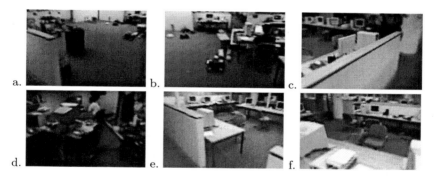

Fig. 13. Views generated by the system as the user conducts a virtual tour of the library.

in Figure 12a, Figure 12b shows the result of applying pose estimation and interpolation to the rest of the video sequence. Figure 13 shows some samples of images created as the user explores this environment interactively. Notice that the user can freely enter and exit various rooms and alcoves in the laboratory.

Figure 5 shows the results of applying the reconstruction procedure to 14 images acquired from a sequence taken inside an abandoned hospital building. This figure demonstrates the capability of constructing polyhedral models from the recovered model features.

The fact that the reconstruction process can be carried out entirely from the video sequence simplifies the process of data collection. Figure 14c shows a mobile platform outfitted with an omnidirectional camera system produced by Remote Reality inc.. This system was used to acquire the imagery that was used to construct the model shown in Figure 5. Note that the only sensor carried by this robot is the omnidirectional camera it does not have any odometry or range sensors. During the data collection process the system was piloted by a remote operator using an RC link.

The video data that was used to construct the models shown in Figures 8 and 12 was collected with a handheld omnidirectional camera system as shown in Figure 14. In both cases the video data was captured on a Sony Digital camcorder and transferred to a PC for processing using an IEEE 1394 Firewire link. The images were digitized at a resolution of 720x480 at 24 bits per pixel.

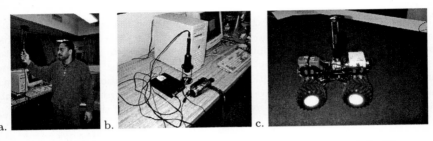

Fig. 14. a. The video imagery used to produce the reconstructions of the library and the laboratory environments was acquired using a handheld omnidirectional camera system b. The equipment used to acquire the data c. Mobile platform equipped with an omnidirectional camera system that was used to acquire video imagery of an indoor environment.

4 Related Work

The idea of using omnidirectional camera system for reconstructing environments from video imagery in the context of robotic applications has been explored by Yagi, Kawato, Tsuji and Ishiguro [21,8,7,9]. These authors presented an omnidirectional camera system based on a conical mirror and described how the measurements obtained from the video imagery acquired with their camera system could be combined with odometry measurements from the robot platform to construct maps of the robots environment. The techniques described in this paper do not require odometry information which means that they can be employed on simpler platforms like the one shown in Figure 14c which are not equipped with odometers. It also simplifies the data acquisition process since we do not have to calibrate the relationship between the camera system the robots odometry system.

Szeliski and Shum [18] describe an interactive approach to reconstructing scenes from panoramic imagery which is constructed by stitching together video frames that are acquired as a camera is spun around its center of projection.

Coorg and Teller [3] describe a system which is able to automatically extract building models from a data set of panoramic images augmented with pose information which they refer to as pose imagery

From the point of view of robotic applications, reconstruction techniques based on omnidirectional imagery are more attractive than those that involve constructing panoramas from standard video imagery since they do not involve moving the camera and since the omnidirectional imagery can be acquired as the robot moves through the environment.

The process of acquiring omnidirectional video imagery of an immersive environment is much simpler than the process of acquiring panoramic images. One would not really consider constructing a sequence of tightly spaced panoramic images of an environment because of the time required to acquire the imagery and stitch it together. However, this is precisely the type of data contained in an omnidirectional video sequence. By estimating the pose at every location in the sequence the Video Plus system is able to fully exploit the range of viewpoints represented in the image sequence.

Boult [1] describes an interesting system which allows a user to experience remote environments by viewing video imagery acquired with an omnidirectional camera. During playback the user can control the direction from which she views the scene interactively. The VideoPlus system described in this paper provides the end user with the ability to control her viewing position as well as her viewing direction. This flexibility is made possible by the fact that the video imagery is augmented with pose information which allows the user to navigate the sequence in an order that is completely different from the temporal ordering of the original sequence.

The VideoPlus system in similar in spirit to the Movie Map system described by Lippman [13] and to the QuickTime VR system developed by Chen [2] in that the end result of the analysis is a set of omnidirectional images annotated with position. The user is able to navigate through the scene by jumping from one image to another. The contribution of this work is to propose a simple and effective way of recovering the positions of the omnidirectional views from image measurements without having to place artificial fiducials in the environment or requiring a separate pose estimation system.

Shum and He [16] describe an innovative approach to generating novel views of an environment based on a set of images acquired while the camera is rotated around a set of concentric circles. This system builds on the plenoptic sampling ideas described by Levoy and Hanrahan [11] and Gortler, Grzeszczuk, Szeliski and Cohen [6]. The presented approach shares the advantage of these image based rendering techniques since the VideoPlus scheme allows you to explore arbitrarily complex environments without having to model the geometric and photometric properties of all of the surfaces in the scene. The rerendered images are essentially resampled versions of the original imagery. However, the scheme presented in this paper dispenses with the need for a specific camera trajectory and it can be used to capture the appearance of extended environments such as office complexes which involve walls and other occluding features which are not accounted for by these plenoptic sampling schemes.

5 Future Work

The scheme used to generate views of an environment during a walkthrough is currently quite simple. Given the users desired viewpoint the system selects the omnidirectional image that is closest to that location and generates an image with the appropriate viewing direction. The obvious limitation of this approach is that the viewing position is restricted to locations which were imaged in the original video sequence.

This limitation can be removed by applying image based rendering techniques. One approach to generating novel images is to resample the intensity data from other images depending on the hypothesized structure of the scene. The video plus system has access to the positions of all of the frames in the sequence along with a coarse polyhedral model of the environment which could be used to transfer pixel data from the original views to the virtual view.

Another approach to generating novel views would be to find correspondences between salient image features in nearby omnidirectional images in the sequence and to use these correspondences to construct a warping function which would map pixels from the original images to the virtual viewpoint [12].

The success of any view generation technique will depend upon having a set of images taken from a sufficiently representative set of viewpoints. A better understanding of how to go about capturing such a data set taking into account the structure of the scene and the viewpoints that are likely to be of most interest is needed. The ultimate goal would be to produce a system where the user could arbitrarily select the desired viewpoint and viewing direction so as to explore the environment in an unconstrained manner.

The largest drawbacks to using omnidirectional video imagery is the reduced image resolution. This effect can be mitigated by employing higher resolution video cameras. One of the tradeoffs that is currently being explored is the possibility of acquiring higher resolution imagery at a lower frame rate. This would allow us to produce sharper images of the scene but would either slow down the data acquisition process or require better interpolation strategies.

6 Conclusions

This paper presents a simple approach to capturing the appearance of immersive scenes based on an omnidirectional video sequence. The system proceeds by combining techniques from structure from motion with ideas from image based rendering. An interactive photogrammetric modeling scheme is used to recover the positions of a set of salient features in the scene (points and lines) from a small set of keyframe images. These features are then used as fiducials to estimate the position and orientation of the omnidirectional camera at every frame in the video clip.

By augmenting the video sequence with pose information we provide the end user with the capability of indexing the video sequence spatially as opposed to temporally. This means that the user can explore the image sequence in ways that were not envisioned when the sequence was initially collected.

The cost of augmenting the video sequence with pose information is very slight since it only involves storing six numbers per frame. The hardware requirements of the proposed scheme are also quite modest since the reconstruction is performed entirely from the image data. It does not involve a specific camera trajectory or a separate sensor for measuring the camera position. As such, the method is particularly appropriate for immersive man-made structures where GPS data is often unavailable.

We envision that this system could be used to acquire representations of immersive environments, like museums, that users could then explore interactively. It might also be appropriate for acquiring immersive backgrounds for video games or training simulators.

Future work will address the problem of generating imagery from novel viewpoints and improving the resolution of the imagery generated by the system.

Acknowledgements. This research was supported by the National Science Foundation under a CAREER grant (IIS98-7687).

References

1. Terrance E. Boult. Remote reality via omni-directional imaging. In Scott Grisson, Janet McAndless, Omar Ahmad, Christopher Stapleton, Adele Newton, Celia Pearce, Ryan Ulyate, and Rick Parent, editors, *Conference abstracts and applications: SIGGRAPH 98, July 14-21, 1998, Orlando, FL*, Computer Graphics, pages 253-253, New York, NY 10036, USA, 1998. ACM Press.
2. S. E. Chen. Quicktime vr - an image-based approach to virtual environment navigation. In *SIGGRAPH*, pages 29-38, August 1995.
3. Satyan Coorg and Seth Teller. Automatic extraction of textured vertical facades from pose imagery. Technical report, MIT Computer Graphics Group, January 1998.
4. Paul E. Debevec, Camillo J. Taylor, and Jitendra Malik. Modeling and rendering architecture from photographs: A hybrid geometry- and image-based approach. In *Proceedings of SIGGRAPH 96. In Computer Graphics Proceedings, Annual Conference Series*, pages 11-21, New Orleans, LA, August 4-9 1996. ACM SIGGRAPH.
5. C. Geyer and K. Daniilidis. Catadioptric camera calibration. In *International Conference on Computer Vision*, pages 398-404, 1999.
6. Steven J. Gortler, Radek Grzeszczuk, Richard Szeliski, and Michael Cohen. The lumigraph. In *Proceedings of SIGGRAPH 96. In Computer Graphics Proceedings, Annual Conference Series*, pages 31-43, New Orleans, LA, August 4-9 1996. ACM SIGGRAPH.
7. Hiroshi Ishiguro, Takeshi Maeda, Takahiro Miyashita, and Saburo Tsuji. A strategy for acquiring an environmental model with panoramic sensing by a mobile robot. In *IEEE Int. Conf. on Robotics and Automation*, pages 724-729, 1994.

8. Hiroshi Ishiguro, Kenji Ueda, and Saburo Tsuji. Omnidirectional visual information for navigating a mobile robot. In *IEEE Int. Conf. on Robotics and Automation*, pages 799–804, 1993.
9. Hiroshi Ishiguro, Masashi Yamamoto, and Saburo Tsuji. Omni-directional stereo. *IEEE Trans. Pattern Anal. Machine Intell.*, 14(2):257–262, February 1992.
10. J.E. Dennis Jr. and Robert B. Schnabel. *Numerical Methods for Unconstrained Optimization and Nonlinear Equations.* Prentice-Hall, 1983.
11. Marc Levoy and Pat Hanrahan. Light field rendering. In *Proceedings of SIGGRAPH 96. In Computer Graphics Proceedings, Annual Conference Series*, pages 31–43, New Orleans, LA, August 4-9 1996. ACM SIGGRAPH.
12. Maxime Lhuillier and Long Quan. Image interpolation by joint view triangulation. In *Proc. IEEE Conf. on Comp. Vision and Patt. Recog.*, volume 2, pages 139–145, June 1999.
13. A. Lippman. Movie maps: An application of the optical video-disc to computer graphics. In *SIGGRAPH*, pages 32–42, July 1980.
14. B.D. Lucas and T. Kanade. An iterative image registration technique with an application to stereo vision. In *Proc. 7th International Joint Conference on Artificial Intelligence*, 1981.
15. Shree Nayar. Catadioptric omnidirectional camera. In *Proc. IEEE Conf. on Comp. Vision and Patt. Recog.*, 1997.
16. Heung-Yeung Shum and Li-Wei He. Rendering with concentric mosaics. In *SIGGRAPH*, pages 299–306, August 1999.
17. Tomas Svoboda, Tomas Pajdla, and Vaclav Hlavac. Epipolar geometry for panoramic cameras. In *European Conference on Computer Vision*, pages 218–232. Springer, 1998.
18. R. Szeliski and H. Y. Shum. Creating full ciew panoramic image mosaics and texture-mapped models. In *SIGGRAPH*, pages 251–258, August 1997.
19. Camillo J. Taylor and David J. Kriegman. Minimization on the lie group so(3) and related manifolds. Technical Report 9405, Center for Systems Science, Dept. of Electrical Engineering, Yale University, New Haven, CT, April 1994.
20. Camillo J. Taylor and David J. Kriegman. Structure and motion from line segments in multiple images. *IEEE Trans. Pattern Anal. Machine Intell.*, 17(11), November 1995.
21. Yasushi Yagi, Shinjiro Kawato, and Saburo Tsuji. Real-time omnidirectional image sensor (copis) for vision-guided navigation. *IEEE Journal of Robotics and Automation*, 10(1):11–21, February 1994.

Discussion

1. **Marc Pollefeys, K.U.Leuven**: How do you get yourself out of the video?
   C. J. Taylor: Because I choose the camera trajectory, I just don't look at me: I'm behind the camera.
2. **Andrew Fitzgibbon, University of Oxford**: When you say you couldn't automatically determine the camera motion, was that a problem with tracking 2D points?
   C. J. Taylor: Yes and no. I suspect that if I bang a little harder I could get the performance up. It works about 90 to 95 percent of the time but it's sometimes just not good enough. The other issue is of course occlusion. You're walking in and out of rooms, you're walking around and things disappear. The question is how far you want to be able to track and estimate pose. I'd be delighted if somebody gave me a really good industrial strength extended tracker: something that's gonna work if I walk 20 or 30 feet.
3. **Bill Triggs, INRIA Rhône-Alpes**: A lot of us have worked hard on methods to calibrate perspective cameras, which turns out to be quite a delicate problem. Have you noticed that the calibration problem for omni-directional cameras is less severe?
   C. J. Taylor: The nice thing about omni-directional imagery is that they're very easy to calibrate, at least to get a rough calibration out. The trick that Geyer and Daniilidis [1] demonstrated: to use straight lines to improve calibration actually works pretty well. I haven't seen significant calibration problems, yet. Maybe if I try to do much finer, detailed work they may show up.
4. **Kenichi Kanatani, Gunma University**: I am worrying about the resolution inhomogeneity of omni-directional lenses. Don't you think this is a problem?
   C. J. Taylor: Yes, I'm just using essentially an off-the-shelf camera and there is some resolution inhomogeneity. The easy fix is to increase the resolution of your sensor. If you have more pixels, you can do a better job interpolating. There have been some people who have looked at changing mirror geometry and things but for what I want to do, the central projection property is very useful and I don't want to sacrifice that.

References

1. C. Geyer and K. Daniilidis. A unifying theory for central panoramic systems and practical implications. In *Proc. European Conference on Computer Vision*, pages 445–461, 2000.

Eyes from Eyes*

Patrick Baker, Robert Pless, Cornelia Fermüller and Yiannis Aloimonos

Center for Automation Research
University of Maryland
College Park, MD 20742-3275, USA

Abstract. We describe a family of new imaging systems, called Argus eyes, that consist of common video cameras arranged in some network. The system we built consists of six cameras arranged so that they sample different parts of the visual sphere. This system has the capability of very accurately estimating its own 3D motion and consequently estimating shape models from the individual videos. The reason is that inherent ambiguities of confusion between translation and rotation disappear in this case. We provide an algorithm and several experiments using real outdoor or indoor images demonstrating the superiority of the new sensor with regard to 3D motion estimation.

1 Introduction

Technological advances make it possible to arrange video cameras in some configuration, connect them with a high-speed network and collect synchronized video. Such developments open new avenues in many areas, making it possible to address, for the first time, a variety of applications in surveillance and monitoring, graphics and visualization, robotics and augmented reality. But as the need for applications grows, there does not yet exist a clear idea on how to put together many cameras for solving specific problems. That is, the mathematics of multiple-view vision is not yet understood in a way that relates the configuration of the camera network to the task under consideration. Existing approaches treat almost all problems as multiple stereo problems, thus missing important information hidden in the multiple videos. The goal of this paper is to provide the first steps in filling the gap described above. We consider a multi-camera network as a new eye We studied and built one such eye, consisting of cameras which sample parts of the visual sphere, for the purpose of reconstructing models of space. The motivation for this eye stems from a theoretical study analyzing the influence of the field of view on the accuracy of motion estimation and thus in turn shape reconstruction. The exposition continues by first describing the problems of developing models of shape using a common video camera and pointing out inherent difficulties.

In general, when a scene is viewed from two positions, there are two concepts of interest: (a) The 3D transformation relating the two viewpoints. This is a

---

* Patent pending

rigid motion transformation, consisting of a translation and a rotation (six degrees of freedom). When the viewpoints are close together, this transformation is modeled by the 3D motion of the eye (or camera). (b) The 2D transformation relating the pixels in the two images, i.e., a transformation that given a point in the first image maps it onto its corresponding one in the second image (that is, these two points are the projections of the same scene point). When the viewpoints are close together, this transformation amounts to a vector field denoting the velocity of each pixel, called an image motion field. Perfect knowledge of both transformations described above leads to perfect knowledge of models of space, since knowing exactly how the two viewpoints and the images are related provides the exact position of each scene point in space. Thus, a key to the basic problem of building models of space is the recovery of the two transformations described before and any difficulty in building such models can be traced to the difficulty of estimating these two transformations. What are the limitations in achieving this task?

2   Inherent Limitations

Images, for a standard pinhole camera, are formed by central projection on a plane (Figure 1). The focal length is f and the coordinate system $OXYZ$ is attached to the camera, with Z being the optical axis, perpendicular to the image plane.

Scene points $\mathbf{R}$ are projected onto image points $\mathbf{r}$. Let the camera move in a static environment with instantaneous translation $\mathbf{t}$ and instantaneous rotation $\boldsymbol{\omega}$. The image motion field is described by the following equation:

$$\dot{\mathbf{r}} = -\frac{1}{(\mathbf{R} \cdot \hat{\mathbf{z}})} (\hat{\mathbf{z}} \times (\mathbf{t} \times \mathbf{r})) + \frac{1}{f}\hat{\mathbf{z}} \times (\mathbf{r} \times (\boldsymbol{\omega} \times \mathbf{r}))$$

where $\hat{\mathbf{z}}$ is a unit vector in the direction of the Z axis.

There exists a veritable cornucopia of techniques for finding 3D motion from a video sequence. Most techniques are based on minimizing the deviation from the epipolar constraint. In the continuous case the epipolar constraint takes the following form: $(\mathbf{t} \times \mathbf{r}) \cdot (\dot{\mathbf{r}} + \boldsymbol{\omega} \times \mathbf{r}) = 0$ [4].

One is interested in the estimates of translation $\hat{\mathbf{t}}$ and rotation $\hat{\boldsymbol{\omega}}$ which best satisfy the epipolar constraint at every point $\mathbf{r}$ according to some criteria of deviation. Usually the Euclidean norm is considered leading to the minimization of function.[1]

$$E_{ep} = \iint\limits_{image} \left[(\hat{\mathbf{t}} \times \mathbf{r}) \cdot (\dot{\mathbf{r}} + \hat{\boldsymbol{\omega}} \times \mathbf{r})\right]^2 d\mathbf{r} \tag{1}$$

Solving accurately for 3D motion parameters turned out to be a very difficult problem. The main reason for this has to do with the apparent confusion between

---

[1] Other norms (weighted epipolar deviation) have better performance but still suffer from the rotation/translation confusion problem.

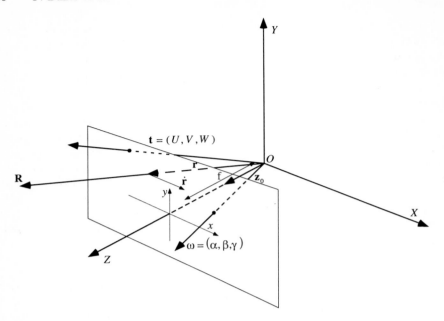

Fig. 1. Image formation on the plane. The system moves with a rigid motion with translational velocity **t** and rotational velocity ω. Scene points **R** project onto image points **r** and the 3D velocity $\dot{\mathbf{R}}$ of a scene point is observed in the image as image velocity $\dot{\mathbf{r}}$.

translation and rotation in the motion field. This is easy to understand at an intuitive level. If we look straight ahead at a shallow scene, whether we rotate around our vertical axis or translate parallel to the scene, the motion field at the center of the image is very similar in the two cases. Thus, for example, translation along the x axis is confused with rotation around the y axis. The basic understanding of this confusion has attracted few investigators over the years [3,4].

Our work is motivated by some recent results analyzing this confusion. In [6, 7] a geometrical statistical analysis of the problem has been conducted. On the basis of (1) the expected value of E_{ep} has been formulated as a five-dimensional function of the motion parameters (two dimensions for $\mathbf{t}/|\mathbf{t}|$ and three for ω). Independent of specific estimators the topographic structure of the surface defined by this function explains the behavior of 3D-motion estimation. Intuitively speaking, it turns out that the minima of this function lie in a valley. This is a cause for inherent instability because, in a real situation, any point on that valley or flat area could serve as the minimum, thus introducing errors in the computation (See Figure 2a).

In particular, the result obtained are as follows: Denote the five unknown motion parameters as (x_0, y_0) (direction of translation) and (α, β, γ) (rotation). Then, if the camera has a limited field of view, *no matter how 3D motion is esti-*

mated from the motion field, the expected solution will contain errors $(x_{0_\epsilon}, y_{0_\epsilon})$, $(\alpha_\epsilon, \beta_\epsilon, \gamma_\epsilon)$ that satisfy two constraints:

(a) The orthogonality constraint: $\dfrac{x_{0_\epsilon}}{y_{0_\epsilon}} = -\dfrac{\beta_\epsilon}{\alpha_\epsilon}$

(b) The line constraint: $\dfrac{x_0}{y_0} = \dfrac{x_{0_\epsilon}}{y_{0_\epsilon}}$

In addition, $\gamma_\epsilon = 0$. The result states that the solution contains errors that are mingled and create a confusion between rotation and translation that cannot be cleared up, with the exception of the rotation around the optical axis (γ). The errors may be small or large, but their expected value will always satisfy the above conditions. Although the 3D-motion estimation approaches described above may provide answers that could be sufficient for various navigation tasks, they cannot be used for deriving object models because the depth Z that is computed will be distorted [2].

The proof in [6,7] is of a statistical nature. Nevertheless, we found experimentally that there were valleys in the function minimized for any indoor or outdoor sequence we worked on. Often we found the valley to be rather wide, but in many cases it was close in position to the predicted one.

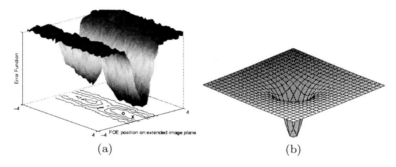

Fig. 2. Schematic illustration of error function in the space of the direction of translation. (a) A valley for a planar surface with limited field of view. (b) Clearly defined minimum for a spherical field of view.

The error function, however, changes as the field of view changes. The remarkable discovery in [6,7] is that when the field of view becomes 360° the ambiguity disappears. This means that there are no more valleys, but a well defined minimum, as shown in Figure 2b. This constitutes the basis of our approach.

Our interest is to develop techniques that, given video data, yield models of the shape of the imaged scene. Since conventional video cameras have an inherent problem, we should perhaps utilize different eyes. If, for example, we had a sensor with a 360° field of view, we should be able to accurately recover 3D motion and subsequently shape. Catadioptric sensors could provide the field of view but they have poor resolution, making it difficult to recover shape models. Thus we built the Argus eye, a construction consisting of six cameras pointing outwards (Figure 3). Clearly, only parts of the sphere are imaged. When this structure is moved arbitrarily in space, then data from all six cameras can be used to very accurately recover 3D motion, which can then be used in the individual videos to recover shape. The next section shows how we calibrated the Argus eye and the final section describes how 3D motion was estimated.

3 Calibration

In order to calibrate the Argus Eye, it is not possible to use ordinary stereo calibration methods, because the fields of view do not overlap. Mechanical calibration is difficult and expensive, so we would like to use an image based method. There are a few possibilities for image based calibration, listed below.

Grid Calibration Construct a precisely measured calibration grid which surrounds the Argus eye, and then use standard calibration methods (such as [8] or [5]) from this. This method is difficult and expensive and we would prefer not to have to implement it.

Self Calibration Use a self-calibration algorithm to obtain the calibration parameters. By matching the axes of rotation in the various cameras, this method should be able to obtain the rotation between the cameras. An estimate of the translation between the cameras would require the computation of depth, which is sensitive to noise, so that it would be difficult to self calibrate the translation between the cameras.

Many Camera Calibration If additional cameras were placed around the Argus eye pointing inwards in such a way that the cones of view of the cameras

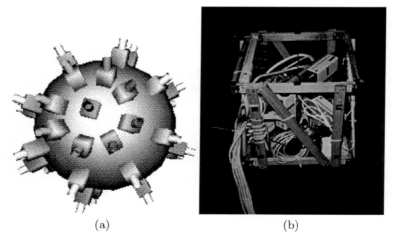

(a) (b)

Fig. 3. (a) A compound-like eye composed of conventional video cameras, and a schematic description of the Argus eye. (b) The actual Argus eye. The cameras are attached to diagonals of a polyhedron made out of wooden sticks.

intersected with each other and with those of the Argus eye, then those cameras, properly calibrated, can be used to calibrate as will be shown. See Figure 4 for a diagram.

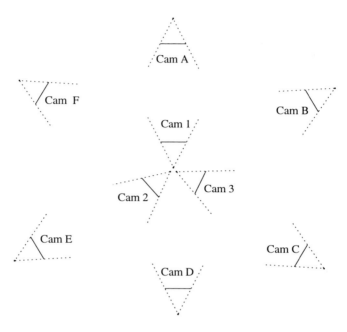

Fig. 4. Arrangement of the cameras for calibration purposes. Cameras 1 to 3 make up an Argus eye. Cameras A-F are auxiliary surrounding cameras.

In order to obtain the most accuracy with the least cost, we decided to implement the solution of calibrating with *auxiliary* cameras, since we have at our disposal a multi-camera laboratory, which has 64 synchronized cameras arranged so that they all point inward toward the center of the room. This idea of using auxiliary cameras, that is, cameras not actually involved in taking the data but integral to the calibration, is an important one. This concept is applicable not only to the task at hand, but to other situations such as stereo calibration, since it can give a much larger depth of field for corresponding points than a grid.

3.1 Method Overview

The Argus eye is calibrated by putting it in the middle of a collection of inward pointing cameras, all synchronized together, as shown in Figure 4. In the darkness an LED wand is waved in a manner so as to fill the cones of view of all the cameras with points. The cameras in this way obtain many accurate point correspondences using simple background subtraction and thresholding to extract the points. The point correspondences, while not between the Argus Eye cameras, serve to transfer the calibration from the auxiliary cameras.

The Argus eye can then be taken away, and a calibration grid put in its place, so that all the inward pointing cameras can be intrinsically and extrinsically calibrated. Using these projection matrices, the actual point location of the LED can be calculated. Since we have the LED world point locations and their projections on the Argus eye, it is as if we had a calibration grid, and we can use any standard algorithm to obtain the projection matrices. From these the rotational matrices are easily obtained using the QR decomposition.

3.2 Method Specifics

There are two complications to implementing this method in a straightforward manner. First, the cameras need to be radially calibrated. Second, the way that the Argus Eye is constructed makes it impossible to place the device in a way so that more than four cameras can see a significant distance at any one time.

Radial calibration. The lenses we use cause a radial distortion of about four pixels on the edges of the image, which needs be corrected. We use this model of radial calibration, where x_n and y_n are the coordinates corrected from the measured x_m and y_m.

$$x_n = (1 + \kappa((x_m - x_c)^2 + (y_m - y_c)^2))(x_m - x_c) + x_c \qquad (2)$$

$$y_n = (1 + \kappa((x_m - x_c)^2 + (y_m - y_c)^2))(y_m - y_c) + y_c \qquad (3)$$

where (x_c, y_c) is the center of radial calibration. In our 644×484 images, we measured the κ parameter at approximately $1e-7$ for our lenses and the center of radial calibration at the approximate image center.

The cameras were radially calibrated with a grid pattern shown to every camera. The points of intersection were extracted from the grid images, and are made homogeneous with $\mathbf{r} = [x_n, y_n, f]$. Then for every set of three points $\mathbf{r}_{i,1}, \mathbf{r}_{i,2}, \mathbf{r}_{i,3}$ which are supposed to be collinear, we use the triple product as error measure:

$$E_i = (\mathbf{r}_{i,1} \times \mathbf{r}_{i,2}) \cdot \mathbf{r}_{i,3}$$

We minimize:

$$\sum_i E_i$$

over κ and x_c, y_c. The results of this radial calibration were very satisfactory, and no other calibration (such as tangential) was necessary.

Room calibration. While we could obtain some calibration by using the calibration frame, it would be more desirable to use all the point correspondences to obtain an accurate calibration throughout the whole room. To begin with, we used a measured calibration frame to obtain projection matrices using the following algorithm.

To start, we use a standard technique in which a calibration frame with known points is set up in the middle of the room. The known world points together with known image points (located by the user) are used in a nonlinear minimization to find the projection matrices. If the world points in the fiducial coordinate system are $\mathbf{R}_1, \mathbf{R}_2, \ldots, \mathbf{R}_N$, and the hypothesized projection matrix is P, then the world point $\mathbf{R}_i$ should project to

$$\mathbf{r}_i = P \begin{bmatrix} \mathbf{R}_i \\ 1 \end{bmatrix} \tag{4}$$

If the measured image points are $\hat{\mathbf{r}}_1, \hat{\mathbf{r}}_2, \ldots, \hat{\mathbf{r}}_N$, then we can measure the error in the projection matrix by the following sum:

$$\sum_i \left(\frac{\hat{\mathbf{r}}_i}{\hat{r}_{i,3}} - \frac{\hat{\mathbf{r}}_i}{r_{i,3}} \right)^2 \tag{5}$$

A nonlinear minimization run on this error measure is sufficient to find the projection matrices.

However, this calibration uses a small number of white balls which are difficult to localize properly. Note that it is necessary to use a frame rather than a grid with more easily localizable line intersections, because the world points must be visible from all directions. Thus, we would like to improve this calibration significantly. We can do this by using the LED data which is, after all, a collection of point correspondences, though we don't know where the world points themselves are.

Given the projection matrices obtained from the calibration frame, we can reconstruct world points given the point correspondences. These calculated world points can then be used as a calibration frame to compute the projection matrices as above.

Argus calibration. As we stated earlier, the calibration was made trickier by the fact that the Argus Eye could not be placed so that all cameras had a significant depth of field. This is because the cameras point in six directions, so at any time, one of them is pointing predominantly towards the floor. Thus the Argus system needed to be calibrated multiple times in order to obtain the most accurate calibration. More specifically, if the Argus Eye is rotated three times, then we need to form an error measure over all six cameras plus the two displacements of the whole Argus Eye from the initial position.

Let the $\mathbf{R}_i, \mathbf{R}_j, \mathbf{R}_k$ be the world points generated in positions 1, 2, and 3, respectively. Then we can form:

$$\mathbf{r}_i = P_l \begin{bmatrix} \mathbf{R}_i \\ 1 \end{bmatrix} \tag{6}$$

$$\mathbf{r}_j = P_l \begin{bmatrix} Q_2 T_2 \\ 0^T 1 \end{bmatrix} \begin{bmatrix} \mathbf{R}_j \\ 1 \end{bmatrix} \tag{7}$$

$$\mathbf{r}_k = P_l \begin{bmatrix} Q_3 T_3 \\ 0^T 1 \end{bmatrix} \begin{bmatrix} \mathbf{R}_k \\ 1 \end{bmatrix} \tag{8}$$

where the Q_i and the T_i specify the rotation and translation in the displacement to the two Argus Eye positions. Then we can optimize over the Q_2, Q_3, T_2, T_3 and P_l as specified above, and throw away the Q's and T's. We are left with just the P_l, which are the projection matrices for the cameras in the Argus system. These projection matrices can then be used in our egomotion algorithms.

4   3D Motion from the Argus Eye

Consider a calibrated Argus eye moving in an unrestricted manner in space, collecting synchronized video from each of the video cameras. We would like to find the 3D motion of the whole system. Given that motion and the calibration, we can then determine the motion of each individual camera so that we can reconstruct shape.

An important aspect of the results in [6,7] is that they are algorithm independent. Simply put, whatever the objective function one minimizes, the minima will lie along valleys. The data is not sufficient to disambiguate further. Let us look at pictures of these ambiguities. We constructed a six camera Argus Eye and processed each of the six sequences with state-of-the-art algorithms in motion analysis. Figure 5 shows (on the sphere of possible translations) the valleys obtained when minimizing the epipolar deviation. Noting that the red areas are all the points within a small percentage of the minimum, we can see clearly the ambiguity theoretically shown in the proofs. Our translation could be anywhere in the red area. With such a strong ambiguity, it's no wonder that shape models are difficult to construct.

We now show how to use the images from *all* the cameras in order to resolve these ambiguities. Let us assume that for every camera i, its projection matrix is:

$$P_i = R_i^T [K_i \mid -\mathbf{c}_i] \tag{9}$$

where K_i is the calibration matrix, $\mathbf{c}_i$ is the position of the camera, and R_i is the rotation of the camera.

Note that when cameras are mounted rigidly together, and if the rotation of each camera i is $\boldsymbol{\omega}_i$, then the rotation $\boldsymbol{\omega} = R_i^T \boldsymbol{\omega}_i$ should be the same for every camera. Now for each camera i, let us consider the set of translations $\{\mathbf{t}_{i,j}\}$ with error close to the minimum. Given a translational estimate $\mathbf{t}_{i,j}$, it is easy to estimate the rotation $\boldsymbol{\omega}_{i,j}$ of the camera using any of a variety of

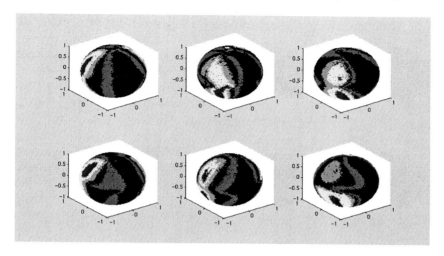

Fig. 5. Valleys obtained on the sphere of possible translations when minimizing the epipolar deviation.

techniques. Here we use a technique by Brodsky in [1], using the minimization of the depth variability constraint: call it $f : \mathbf{T} \to \mathbf{\Omega}$, a function from the set of translations to the set of rotations. This function f is a diffeomorphism, so that given the 2D manifold of candidate translations (the ones with low error), we have a 2D manifold of candidate rotations, which we can derotate by R_i^{T}, to obtain a 2D manifold of rotational estimates in the fiducial coordinate system. Significantly, the *rotations exist in 3D space*, so that from six cameras, we have six 2D manifolds of candidate rotations in the 3D space of possible rotations. We can then find their intersection, which in general should be a single point.

This video confirms two basic tenets of this paper. First, it shows that the motion estimates of lowest error in individual cameras are *not* the correct motions, since if they were, the lowest error points would be coincident in rotation space. Thus even though we are using state-of-the-art algorithms, it is not possible to extract the correct motion from a single camera with limited field of view, as is shown in the proof. Second, the video shows that if we look at *all* the motion candidates of low error, the correct motion is in that set, shown by the intersection of the six manifolds at a single point.

That the manifolds intersect so closely shows we can find the rotation well. Since we know the rotation, the translation is much easier to find. Let us first look at what the translations are in each camera. Given this accurate rotation, the translational ambiguity is confined to a very thin valley, shown in Figure 6. If we can find a way to intersect the translations represented by these valleys, then we can find the complete 3D translation.

We must look more closely at how the fiducial motion is related to the individual camera motion. Each camera's translation consists of the translation of the system added to the translation due to the rotation of the whole system

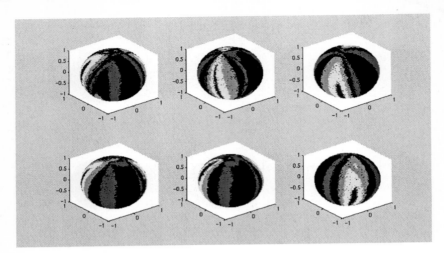

Fig. 6. Valleys obtained on the sphere of possible translations once an accurate estimate of the rotation is obtained.

crossed with the camera position.

$$\mathbf{t}_i = R_i(\mathbf{t} + \boldsymbol{\omega} \times \mathbf{c}_i) \qquad (10)$$

We need to search the 3D space of possible system translations to minimize the sum of the epipolar deviations from all cameras using the translation derived from the above equation and the rotation derived earlier. In Figure 7 we see the location of the low error translations in a spherical slice 3D translation space. Notice the well-defined minimum (in red), indicating the direction of the translation obtained is not ambiguous, so that a minimization procedure like Levenberg-Marquardt will be able to find a unique minimum for the direction of translation.

Looking at (10), we notice something interesting, if not necessarily surprising. If the fiducial translation were 0, then each camera translation would be completely a function of the calibration and the rotation $\boldsymbol{\omega}$. But since we know the rotation exactly, we can know the translation in each camera *without the scale ambiguity*. In the case where the translation is not much larger than the rotation and the distance between the cameras is significant, it is possible to calculate the absolute translation. Thus camera construction techniques which force the centers of projection to be coincident may have simpler algorithms, but the data is not as rich. Here we can obtain metric depth without using stereo techniques.

The preceding discussion showed how, by utilizing all video sequences, a very accurate estimate for the 3D motion can be obtained. This motion can now be utilized to obtain shape models. Figure 8 shows a sequence taken by one of the cameras of the Argus eye. By utilizing all six videos an estimate of the

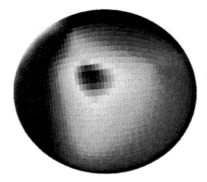

Fig. 7. Location of low error translations in a spherical slice 3D translation space.

3D motion is computed from two frames in the sequence. Figure 9 shows the recovered model.

Fig. 8. A few frames from a sequence taken by one of the cameras of the argus eye

5 Conclusions

Our work is based on recent theoretical results in the literature that established the robustness of 3D motion estimation as a function of the field of view. We built a new imaging system, called the Argus eye, consisting of six high-resolution cameras sampling a part of the plenoptic function. We calibrated the system and developed an algorithm for recovering the system's 3D motion by processing all synchronized videos. Our solution provides remarkably accurate results that can be used for building models from video, for use in a variety of applications.

References

1. T. Brodský, C. Fermüller, and Y. Aloimonos. Structure from motion: Beyond the epipolar constraint. *International Journal of Computer Vision*, 37:231–258, 2000.
2. L. Cheong, C. Fermüller, and Y. Aloimonos. Effects of errors in the viewing geometry on shape estimation. *Computer Vision and Image Understanding*, 71:356–372, 1998.

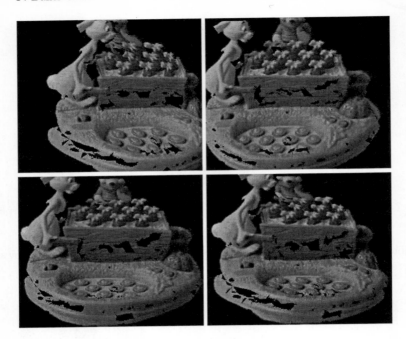

Fig. 9. Some views of the recovered model.

3. K. Daniilidis. *On the Error Sensitivity in the Recovery of Object Descriptions.* PhD thesis, Department of Informatics, University of Karlsruhe, Germany, 1992. In German.
4. K. Daniilidis and M. E. Spetsakis. Understanding noise sensitivity in structure from motion. In Y. Aloimonos, editor, *Visual Navigation: From Biological Systems to Unmanned Ground Vehicles*, Advances in Computer Vision, chapter 4. Lawrence Erlbaum Associates, Mahwah, NJ, 1997.
5. O. Faugeras and G. Toscani. The calibration problem for stereo. In *Proc. IEEE Conference on Computer Vision and Pattern Recognition*, pages 364–374, Miami Beach, FL, 1986.
6. C. Fermüller and Y. Aloimonos. Ambiguity in structure from motion: Sphere versus plane. *International Journal of Computer Vision*, 28:137–154, 1998.
7. C. Fermüller and Y. Aloimonos. Observability of 3D motion. *International Journal of Computer Vision*, 37:43–63, 2000.
8. R. Y. Tsai. An efficient and accurate camera calibration technique for 3D machine vision. In *Proc. IEEE Conference on Computer Vision and Pattern Recognition*, pages 364–374, Miami Beach, FL, 1986.

Discussion

1. **Luc Van Gool, Katholieke Universiteit Leuven**: You take 6 images now instead of only having 1. In that sense you lose a lot of the available resolution which is not focussed on the object of interest but rather on the

environment to have a more stable estimation of the camera motion. But is the comparison really fair in the sense that the sequence of the leopard you showed is actually very short. It is almost close to a pure translation which is close to a degenerate case anyway. Normally you would rather move around the object, have more different views from a wider gamut of orientations. Is that still not a better alternative, because then the image information is used to a fuller extent for the real reconstruction?

Tomas Brodsky: Yes, certainly. Maybe the sequence of the leopard is not the best example to show the power of this. If you take it in a room, then even though the views see different parts, they basically give you information about the same geometry. You could patch them together. I think we'll see better results as time progresses because this is really a fairly new device. Again, this is the difference: if you do point correspondences you have to work on matching but then you easily get wide baseline views so your structure estimates tend to be better. Here, if you look at ten sequences you still have to work on linking many frames together.

2. **Bill Triggs, INRIA Rhône-Alpes**: A comment and then a question. My intuition suggests that the ill-conditioning is caused by the rotation versus translation ambiguity. If your scene is a small compact object with little relief, there is little residual parallax, so it is difficult to tell tracking from panning (sideways translation from sideways rotation). But another camera at right angles to the first sees a completely different motion in the two cases (*e.g.*, if it is looking in the tracking direction, forward translation versus sideways rotation). So the question is: why use six cameras, when two would have been enough to break the ambiguity ?

Tomas Brodsky: I don't know why 6 cameras. I'm not sure if two or three would have been enough. It's possible that two would be sufficient.

Bill Triggs: You can think of the ambiguity as being caused by not having enough parallax to tell that you have moved sideways, provided that you also rotate to continued to fixate on the same object. But you can only fixate on a single 3D point at a time—any camera looking in another direction will not be fixated, which will break the ambiguity.

Tomas Brodsky: What I like in using 6 cameras is the robustness because then you combine six different inputs. I wasn't really involved in the design of the device, so I'm not sure why they used 6 cameras and not 4 or fewer.

3. **Hans-Helmut Nagel, University of Karlsruhe**: If you capture an entire room, you don't necessarily know in advance where your problem arises. If you have 6 cameras, you then have additional information, even if the two cameras you focus on first will not be the most appropriate in order to disambiguate.

Tomas Brodsky: If I may add, if you look at the theory and have two cameras looking 90 degrees apart, there still might be certain motions which give you problems. You certainly get no ambiguities for a full field of view. If you want to approximate fly-vision, you will want to use as many cameras as you can.

Sequential Localisation and Map-Building in Computer Vision and Robotics

Andrew J. Davison and Nobuyuki Kita

Intelligent Systems Division, Electrotechnical Laboratory
1-1-4 Umezono, Tsukuba, Ibaraki 305–8568, Japan
{davison,nkita}@etl.go.jp

Abstract. Reviewing the important problem of sequential localisation and map-building, we emphasize its genericity and in particular draw parallels between the often divided fields of computer vision and robot navigation. We compare sequential techniques with the batch methodologies currently prevalent in computer vision, and explain the additional challenges presented by real-time constraints which mean that there is still much work to be done in the sequential case, which when solved will lead to impressive and useful applications. In a detailed tutorial on map-building using first-order error propagation, particular attention is drawn to the roles of modelling and an active methodology. Finally, recognising the critical role of software in tackling a generic problem such as this, we announce the distribution of a proven and carefully designed open-source software framework which is intended for use in a wide range of robot and vision applications: **http://www.robots.ox.ac.uk/~ajd/**

1 Introduction

Structure from motion in computer vision and simultaneous map building and localisation for mobile robots are two views of the same problem. The situation under consideration is that of a body which moves through a static environment about which it has little or no prior knowledge, and measurements from its sensors are used to provide information about its motion and the structure of the world. This body could be a single camera, a robot with various sensors, or any other moving object able to sense its surroundings. Nevertheless, in recent years research has taken many paths to solving this problem with a lack of acknowledgement of its general nature, with a particular divide arising between robotic and "pure vision" approaches.

Crucially, in this paper we are interested in the **sequential** case, where map-building and localisation are able to proceed in a step-by-step fashion as movement occurs. We will contrast this with situations where the **batch** methods currently prevalent in computer vision (and their cousins in robot map-building) can be applied, where measurements from different time steps are used in parallel after the event. Despite renewed interest in sequential map-building from the robotics community, in computer vision recent successful work in off-line reconstruction from image sequences has conspicuously not been accompanied by advances in real-time methods. Sequential map-building is a problem which is far from being solved, and we will look at the state of the art and its limitations.

1.1 Applications

In map-building applications where localisation or map estimates are needed quickly and successively, either to supply data to external processes in real-time or to feed back into determining future actions, **only sequential methods can be used**. For instance:

- Camera-based structure from motion methods that need to update in real-time, like in an inside-out head-tracking application, where an outward-looking camera attached to a head-mounted display user's head identifies and tracks arbitrary features in the surroundings to calculate head movement, or in live virtual studio applications, where the movement of television cameras needs to be known precisely so that live and computer-generated images can be fused to form composite output.
- Autonomous robot navigation in unknown environments, where sensor readings are used to build and update maps, and continually estimate the robot's movement through the world.

1.2 Aims of This Paper

1. To review and clarify the status of the sequential map-building problem, and emphasize its genericity within robotics and computer vision.
2. In a detailed tutorial on map-building using first-order error propagation, to discuss a number of details about implementing real sequential systems and explain the approaches our experience has led to.
3. To announce the distribution of an open-source software package for sequential localisation and map-building, designed with a realisation of the general nature of this class of problems and therefore readily applicable in many applications, and already proven in two research projects [7,6],

2 The Challenges of Sequential Map-Building

2.1 The Main Point

Thinking first not of actively adding to a map, but of updating uncertain estimates of the locations of various features and that of a moving sensor platform measuring them in a sequential, real-time sense: **the amount of computation which can be carried out in each time-step is bounded by a constant.** This follows simply from thinking of implementing such a system: however fast the processor available, it can only do so much in a certain time-step.

A major implication of this is that **at a given time, we must express all our knowledge of the evolution of the system up to that time with an** amount of information bounded by a constant. The previous knowledge must be combined with any new information from the current time step to produce updated estimates within the finite processing time available.

In the following sections, we will look at the approach we are forced to take to fit this constraint and the difficulties this presents, since in sequential processing compromises must be made to maintain processing speed.

2.2 Approaching Sequential Map-Building

There would seem to be a difference between robot map-building for localisation, where the goal is to determine the robot's motion making use of arbitrary features in the environment as landmarks, and structure from motion, where the interest is in the scene structure and not in the arbitrary path of the camera used to study it. However, it is necessary, explicitly or implicitly, to estimate both sensor motion and scene structure together if either is to be determined.

Batch Methods: The optimal way to build maps from measurements from a moving robot or sensor is to take all the data obtained over a motion sequence and analyse it **at once**. Estimates of where the robot was at each measurement location are calculated altogether, along with the locations of the features observed, and then adjusted until they best fit all the data. This is the **batch** methodology which is used in state of the art geometrical vision to recover 3D structure maps from video sequences and auto-calibrate cameras (e.g. [20,17]). In robot navigation, batch methods have a shorter history but have appeared recently under the banner **EM** [19], where maps of natural features were formed from a data set collected in an initial guided robot tour of an area; afterwards the robot could use the map during autonomous navigation. However, while batch methods can build optimal maps from previously acquired data sets, they do not offer a way to incrementally change maps in the way required by real-time applications as new data is acquired. The key to why not is that the processing effort needed to calculate an estimate for each robot or camera location on a trajectory depends on the **total number** of locations. If the robot or camera moves to a new location and we wish to combine new measurement data with an existing map, all previous estimates must be revised. This does not fit our requirement for constant processing cost for sequential applications.

A comment on uncalibrated methods, which have become closely intertwined with batch estimation in computer vision: most advanced structure from motion approaches operate by assuming that certain parameters defining the camera's operation (such as the focal length) are completely unknown, and calculations take place in versions of the world which are warped in some unknown way relative to reality via the mathematics of projective geometry. Resolution to real Euclidean estimates only happens as a final step, often after an auto-calibration procedure. It should be remembered that estimating the unknown calibration parameters of a camera in this way is somewhat of a detail when it comes to the general problem of reconstructing the world from uncertain sensor measurements

These warped worlds are of no use when we wish to incorporate other types of information; this could be data from other sensors, but most importantly we mean motion information. Motion models are inextricably tied to the real, Euclidean world — the argument may be made that things such as straight line motion are preserved under projective transformation, but physics is also about rotations and scales (such as that provided by the constant of gravitational acceleration). We argue that the best course for sequential methods is to place estimates in a Euclidean frame straight away. Of course some things may be uncertain, such as absolute scales or calibration parameters, but this uncertainty can be included explicitly in estimates and reduced as more data is acquired. We do not lose any deductive power by doing this: we only add it. As will be explained below, when the interdependence between estimates is propagated properly, we retain the ability to, for instance, estimate the ratio of the depths of two features with a high precision, even if either individually is poorly known.

Doing Things Sequentially: To tackle the sequential case, we need a representation of the current "state" of the system whose size does not vary over time. Then, this state can be updated in a constant time when new information arrives. Both the state and the new information will be accompanied by uncertainty, and we must take account of this when weighting the old and new data to produce updates. We are taken unavoidably into the domain of time-dependent **statistics**, whereas the optimisation approach used in batch methods permits a more lax handling of uncertainty.

Something to clarify early on is that when we talk here about a state being of constant size, we mean that for a map with a given number of features the state size does not change with **time**. The fact that the state size, and therefore processing burden, will increase as the number of features grows seems unavoidable. So to process maps in real-time, we will be limited to a finite number of features. How to deal with more features than this limit is the main challenge of sequential map-building research, and we will look at this further in Section 2.4.

Many early authors [8,11,1] took simple approaches to representing the state and its uncertainty; the locations of the moving robot in the world and features were stored and updated independently. However, if any type of long-term motion is attempted, these methods prove to be deficient: though they produce good estimates of instantaneous motion, they do not take account of the interdependence of the estimates of different quantities, and maps are seen to drift away from ground truth in a systematic way, as can be seen in the experiments of the authors referenced above. They are not able to produce sensible estimates for long runs where previously seen features may be revisited after periods of neglect, an action that allows drifting estimates to be corrected [7].

To give a flavour of the interdependence of estimates in sequential map-building, and emphasize that it is important to estimate robot and feature positions together, steps from a simple scenario are depicted in Figure 1. The sequence of robot behaviour here is not intended to be optimal; the point is

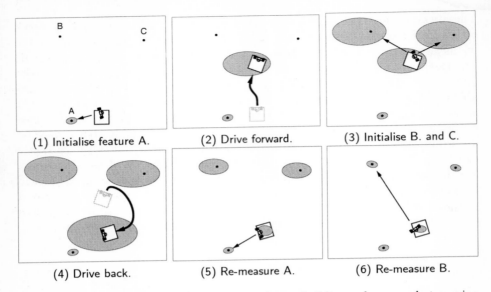

Fig. 1. Six steps in a example of sequential map-building, where a robot moving in two dimensions is assumed to have a fairly accurate sensor allowing it to detect the relative location of point features, and less accurate odometry for dead-reckoning motion estimation. Black points are the true locations of environmental features, and grey areas represent uncertain estimates of the feature and robot positions.

that a map-building algorithm should be able to cope with arbitrary actions and make use of all the information it obtains.

In (1), a robot is dropped into an environment of which it has no prior knowledge. Defining a coordinate frame at this starting position, it uses a sensor to identify feature A and measure its position. The sensor is quite accurate, but there is some uncertainty in this measurement which transposes into the small grey area representing the uncertainty in the estimate of the feature's position.

The robot drives forward in (2), during this time making an estimate of its motion using dead-reckoning (for instance counting the turns of its wheels). This type of motion estimation is notoriously inaccurate and causes motion uncertainties which grow without bound over time, and this is reflected in the large uncertainty region around the robot representing its estimate of its position. In (3), the robot makes initial measurements of features B and C. Since the robot's own position estimate is uncertain at this time, its estimates of the locations of B and C have large uncertainty regions, equivalent to the robot position uncertainty plus the smaller sensor measurement uncertainty. However, although it cannot be represented in the diagram, the estimates in the locations of the robot, B and C are all **coupled** at this point. Their relative positions are quite well known; what is uncertain is the position of the group as a whole.

The robot turns and drives back to near its starting position in (4). During this motion its estimate of its own position, again updated with dead-reckoning, grows even more uncertain. In (5) though, re-measuring feature A, whose absolute location is well known, allows the robot dramatically to improve its position estimate. The important thing to notice is that this measurement **also improves the estimate of the locations of features B and C**. Although the robot had driven farther since first measuring them, estimates of these feature positions were still partially coupled to the robot state, so improving the robot estimate also upgrades the feature estimates. The feature estimates are further improved in (6), where the robot directly re-measures feature B. This measurement, while of course improving the estimate of B, also improves C due to their interdependence (the relative locations of B and C are well known).

At this stage, all estimates are quite good and the robot has built a useful map. It is important to understand that this has happened with a small number of measurements because use has been made of the coupling between estimates.

2.3 Propagating Coupled Estimates

To work with coupled estimates, it is necessary to propagate not only each estimated quantity and its uncertainty, but also how this relates to the uncertainties of other estimates. Generally, a group of uncertain quantities is represented by a probability distribution in multiple dimensions, the form of which will depend on the specific agents of uncertainty in the system. Representing arbitrary probability distributions is not straightforward: one approach uses many random samples (sometimes called particles) to build up the shape, and has recently successfully been seen in vision in the form of the Condensation algorithm [12] for robust contour tracking. This approach has also been used in robot navigation for the problem of localisation using a known map [10], performing extremely well even for the difficult problem of re-localising a robot which is completely lost. However, these Monte Carlo methods are computationally expensive, and particularly are not applicable to the very high-dimensional map-building problem, since the number of particles N, and therefore the computational burden, needed to represent fairly a probability distribution in dimension d varies as:

$$N \geq \frac{D_{min}}{\alpha^d} ,$$

where D_{min} and α are constants with $\alpha << 1$. In modern Condensation applications, the number of dimensions under consideration is limited by computing power to perhaps something less than 10 if realtime operation is desired. This is of course sufficient for estimating the location of a robot with a known map, but not when we simultaneously need to estimate map parameters.

Currently more feasible is to propagate first order approximations to probability distributions. Each estimated parameter is accompanied by single numbers representing its variance and covariance with other parameters — a vector of parameters has a covariance matrix filled with these elements. The Kalman Filter is an optimal solution to linear problems in which all noise sources are gaussian in profile; however, most map-building scenarios are not linear so in these cases the Extended Kalman Filter provides an approximation which in general has been found to perform very well. Called "Stochastic Mapping" in its first correctly-formulated application to robot map-building [18], the EKF has been implemented successfully in different scenarios by other researchers [2,4,5,7,9, 14]. Its main weakness compared to the Monte Carlo methods is its inability to represent multi-modal distributions — where an estimate has two or more peak values that are most likely, with unlikely regions in between. The Monte Carlo methods gain greatly in robustness through this ability. In first order approaches, multiple hypotheses must be considered externally and explicitly.

2.4 The Problems

We consider that there are two main challenges in sequential map-building using first order uncertainty — they are both things we have already touched on:

1. The growth of computational complexity with the number of map features.
2. Coping with mismatches (sometimes known as data association).

We will see in Section 3 that correctly propagating the interdependence of map estimates requires a covariance matrix to be updated whose number of elements grows as the square of the number of estimated parameters. Clearly, the processing requirements quickly get out of hand as the map grows.

The first approach which can be taken is to keep the number of features low using sensible map management. With today's processing power, there need be little trouble with maintaining maps with features in the 50–100 range at decent update rates. This is quite sufficient for many localisation tasks within confined areas: the emphasis should be on using a small number of high-quality features, and on **map using** rather than wasteful map building. For the situation in question, the question "how many features need to be measurable at any given time?" can be asked. The answer will depend on the localising power of a single feature measurement, which depends on the sensor and feature types (for instance, for a camera moving in 3D, seeing just a single point feature will not improve estimates along the several degrees of freedom which that measurement does not constrain, whereas for a robot moving in one dimension with a range sensor, measuring one feature tells it all it needs to know), as well as potential desires for redundant measurements to improve robustness (see Section 3.5). A management algorithm can then add new features to the map only in places where less than this desired number is available.

For applications where the number of features must be larger, various authors have looked at ways to relieve the complexity of large maps. One simple approach, similar to the map management above, is to delete features from the map which do not provide much information[9]: for instance, if two features lie close together, and their estimates have become closely coupled, one can be deleted without sacrificing much information. This opens up the question of which features provide the best localisation information — something also looked at in [5] with regard to active choice among candidate measurements.

Other approaches split a large map into sub-maps, within each of which fully-coupled map-building goes on as normal but between which full coupling is not maintained [4,15]. While in a certain area, the robot will only observe features in the current submap and only update estimates for this submap in real-time. This is an approximation, since in truth each feature estimate is related to every other in the world map, but can be effective when the coupling between submaps can be represented as a single parameter: all the estimates in submap A have a similar dependence on all of those in submap B. This is likely to be the case if the robot spends long periods of time confined to each submap region, rather than frequently moving between them. There are still many issues to settle with these approaches, such as how submap areas should be delineated.

McLauchlan's VSDF [16] is a powerful framework which marries sequential and batch methods, and has been used in several different vision applications. It is based on the propagation of inverse covariance matrices (called information matrices), a strategy which provides some computational advantages, and offers an efficient sequential mode, though this mode makes implicit approximations by ignoring some non-diagonal matrix elements and it is not clear how these approximations compare with other possibilities. Covariance Intersection [23], a general tool for distributed estimation, has also been touted as suitable for map-building, though its generalisations may be too forgiving to produce estimates as good as other more specific methods.

A final approach [5] suggests that while maintaining a full single map, current active parts can be kept fully up to date while currently uninteresting regions are kept on "the back burner" — or perhaps in a hierarchy of updatedness — and proves that it is possible to bring these back up to date at a later stage with no loss of information. This method is certainly interesting, particularly because the hierarchy idea provides possibilities for multiple-hypothesis branching at various levels, but presents large challenges in terms of management: deciding which parts should be updated or left to simmer.

We will look in Section 3.5 at how it is possible to use redundancy and methods like RANSAC to reduce the chance of falling prey to mismatches at the measurement stage, but more generally dealing with this problem requires a multiple-hypothesis approach where estimates fork and the decision on which branch is correct is postponed until more evidence is available. There have yet to be any convincing demonstrations of how this can be incorporated into rigorous sequential map-building. This should be a major focus of future research, since there is no point in improving efficiency with methods which are still prone to instant failure at a mismatch. In our experience, map-building of the type described in Section 3 often can survive a mismatch, though this is by luck since the method includes no model of these events.

3 Map-Building with First Order Uncertainty Propagation: Details and Insights

In the following section we will look in detail at sequential map-building using first-order uncertainty propagation. On its own this represents an obvious and rigorous approach to map-building, but it is also the backbone of the methods described in the previous section for improving efficiency. We will refer throughout to the moving sensor body as "the robot", but this can apply to a single camera or other unrobotlike object. "Feature" is also a general term referring to any object the robot is capable of observing. Features can be points, lines, planes, cylinders or any other type of geometrical object.

3.1 The State Vector and Its Covariance

Current estimates of the state of the robot and the scene features which are known about are stored in the system state vector $\hat{\mathbf{x}}$, and the uncertainty of the estimates in the covariance matrix P. $\hat{\mathbf{x}}$ and P will change in size dynamically as features are added to or deleted from the map. They are partitioned as follows:

$$\hat{\mathbf{x}} = \begin{pmatrix} \hat{\mathbf{x}}_v \\ \hat{\mathbf{y}}_1 \\ \hat{\mathbf{y}}_2 \\ \vdots \end{pmatrix} \quad , \quad P = \begin{bmatrix} P_{xx} & P_{xy_1} & P_{xy_2} & \cdots \\ P_{y_1 x} & P_{y_1 y_1} & P_{y_1 y_2} & \cdots \\ P_{y_2 x} & P_{y_2 y_1} & P_{y_2 y_2} & \cdots \\ \vdots & \vdots & \vdots & \end{bmatrix}.$$

$\hat{\mathbf{x}}_v$ is the robot state estimate, and $\hat{\mathbf{y}}_i$ the estimated state of the ith feature. By the "state" of the robot and features, generally we mean a vector of all the parameters in which we are interested relating to those objects. Of course this means their positions, which can be defined by a number of parameters depending on the geometrical type of the object and dimensionality of the map; but also, there may be other parameters which we would like the estimate, usually because they will affect future motion or measurements.

For dynamically moving objects it is necessary to estimate higher-order motion parameters (velocity, acceleration, etc.). The number of derivatives needed depends on the expected motion (see Section 3.3 about Motion). As another example, a robot may have redundant axes of movement whose status is important but which are not uniquely defined by the robot's geometrical position. These extra parameters can also be used for **calibration constants**, which are initially only approximately known but whose accuracy will improve with the evolution of the system.

3.2 Coordinate Frames and Initialisation

When a robot moves in surroundings which are initially completely unknown, the choice of a world coordinate frame is arbitrary. The only things that can be reported on are the location of the robot relative to any features detected. Indeed, one possible approach is to do away with a world coordinate frame altogether and estimate just the locations of features in a frame which it fixed to the robot; robot motions appear simply as backward feature movements. However, there is not a large computational penalty in including an explicit robot state, and more importantly in most applications of map-building there will be some interaction with information from other sources, which could be in the form of some prior-known feature positions, or maybe simply metric way-points through which the robot is required to move — a world coordinate frame is necessary to interact with information of this kind.

If there is no prior knowledge of the environment, the coordinate frame can be defined to have its origin at the robot's starting position, and the initial uncertainty relating to the robot's position in $\mathbf{P}_{xx}$ is set to zero. If there is prior knowledge of some feature locations, this is put into the map explicitly and it is this which defines the coordinate frame. The robot's starting position relative to these features must also be input, and both robot and feature positions should be assigned suitable initial covariance values. It is not reasonable to set both robot and feature covariances to zero, because their relative locations can never be perfectly known; a typical initial situation would be to have several very well known feature positions with low covariance effectively pinning down the coordinate frame, with a more uncertain robot starting location.

3.3 Motion

What happens to the estimate of the robot's position during a movement? The answer is that we should model its movement as well as we can with a motion model $\mathbf{f}_v(\mathbf{x}_v, \mathbf{u})$, and add to the system covariance to account for our uncertainty in this motion estimate (the process noise $\mathbf{Q}$).

In batch structure from motion, there is typically no motion modelling. The assumption made is that at each new camera position, there is no prior location knowledge; that is to say there is infinite uncertainty (though there may be constraints on some movement dimensions in certain scenarios). In the quasi-static case that these methods are applied to this is sensible. However, when working in the time domain there is always extra information to be had by modelling motion. This model may be very simple or vague, but the best thing to do is to set it up as honestly as possible and make use of it. Quoting from Torr et al. [21], who in turn cite Jaynes [13]:

> Some will complain that to use Bayesian methods one must introduce arbitrary priors on the parameters. However, far from being a disadvantage, this is a tremendous advantage as it forces open acknowledgement of what assumptions were used in designing the algorithm, which all too often are hidden away beneath the veneer of equations.

There are many types of motion model depending on the level of our knowledge about the system. In the case that we have knowledge of the control parameters of a robot (such as "drive forward at 1ms^{-1} for one second with a steering angle of $5°$"), which is the case for a robot which is controlling its own navigation, we can potentially make quite an accurate motion estimate and the process noise covariance will be small. However, if we want to estimate the motion of a camera strapped to an independently manoeuvring human head, we can make much less precise assumptions: for instance, that the head will keep moving more or less at it's current speed, or maybe that it is slightly more likely to slow down than speed up, given that the person is probably moving within a confined space (see the auto-regressive models of [12]).

Process noise accounts for the things that we don't attempt to model. There is of course no such thing as random noise in (classical!) physics: bodies move deterministically. It is just that we can't know all the details of what is happening to them, though in theory we could model everything (slipping wheels; human joints, muscles, chemical reactions!). Models have to stop somewhere: the rest of what is going on we call process noise, and estimate the size of its effect.

The robot's motion is discretised into finite steps, with an incrementing label k affixed to each. There is no need for the length of these steps to be equal in time, though this will often be the case. (One point we would like to highlight is that many navigation researchers have used unnecessarily simple motion models for their mobile robots; e.g. [8], where a model for car-like motion is used which is an approximation for small Δt: in this case it is quite straightforward to construct a motion models which does not require short time steps [5].) In a Jacobian calculation, we change the state and covariance as follows:

$$\hat{\mathbf{x}}(k+1|k) = \begin{pmatrix} \mathbf{f}_v(\mathbf{x}_v(k|k), \mathbf{u}(k)) \\ \hat{\mathbf{y}}_1(k|k) \\ \hat{\mathbf{y}}_2(k|k) \\ \vdots \end{pmatrix} \tag{1}$$

$$P(k+1|k) = \begin{bmatrix} \frac{\partial \mathbf{f}_v}{\partial \mathbf{x}_v} P_{xx}(k|k) \frac{\partial \mathbf{f}_v}{\partial \mathbf{x}_v}^\top + Q(k) & \frac{\partial \mathbf{f}_v}{\partial \mathbf{x}_v} P_{xy_1}(k|k) & \frac{\partial \mathbf{f}_v}{\partial \mathbf{x}_v} P_{xy_2}(k|k) & \cdots \\ P_{y_1 x}(k|k) \frac{\partial \mathbf{f}_v}{\partial \mathbf{x}_v}^\top & P_{y_1 y_1}(k|k) & P_{y_1 y_2}(k|k) & \cdots \\ P_{y_2 x}(k|k) \frac{\partial \mathbf{f}_v}{\partial \mathbf{x}_v}^\top & P_{y_2 y_1}(k|k) & P_{y_2 y_2}(k|k) & \cdots \\ \vdots & \vdots & \vdots & \end{bmatrix} \tag{2}$$

3.4 Measurements: Selection, Prediction and Searching

The way to measure a particular feature i is determined by its feature measurement model $\mathbf{h}_i(\mathbf{x}_v, \mathbf{y}_i)$, and the measurement noise R. Analogous to process noise, measurement noise takes account of the things we don't model in the feature measurement model. Whenever we wish to measure a particular feature, the value of the measurement can be predicted by substituting current estimates $\hat{\mathbf{x}}_v$ and $\hat{\mathbf{y}}_i$ into the expression for $\mathbf{h}_i$. From the predicted value of a measurement, we can calculate, based on knowledge about the particular feature type and saved information on what the first initialisation measurement of this feature was, whether it is worth trying to measure it. For instance, when measuring point features visually with correlation, there is little chance of a successful match if the current viewpoint is far from the original viewpoint. In this way, regions of **measurability** can be defined for each feature, and aid robustness by only allowing match attempts from positions where the chances are good.

The innovation covariance S_i is how much the actual measurement $\mathbf{z}_i$ is expected to deviate from this prediction:

$$S_i = \frac{\partial \mathbf{h}_i}{\partial \mathbf{x}_v} P_{xx} \frac{\partial \mathbf{h}_i}{\partial \mathbf{x}_v}^\top + \frac{\partial \mathbf{h}_i}{\partial \mathbf{x}_v} P_{xy_i} \frac{\partial \mathbf{h}_i}{\partial \mathbf{y}_i}^\top + \frac{\partial \mathbf{h}_i}{\partial \mathbf{y}_i} P_{y_i x} \frac{\partial \mathbf{h}_i}{\partial \mathbf{x}_v}^\top + \frac{\partial \mathbf{h}_i}{\partial \mathbf{y}_i} P_{y_i y_i} \frac{\partial \mathbf{h}_i}{\partial \mathbf{y}_i}^\top + R \ . \quad (3)$$

Calculating S_i before making measurements allows us to form a **search region** in measurement space for each feature, at a chosen number of standard deviations. This is a large advantage because it allows the adoption of an **active** approach: we need only direct searching attention to the this area, maximising computational resources and minimising the chance of obtaining a mismatch.

It is also possible to make decisions on which of several potentially measurable features to observe as a priority based on S_i: if the measurement cost for each candidate is similar, it is favourable to make a measurement where S_i has a high value because there is the most information to be gained here. There is no point in making a measurement when the result is predictable [7]. If measurements are continually chosen like this, the uncertainty in any particular part of the map can be stopped from getting out of control, a situation which would lead to large search regions and the high possibility of mismatches, and to the potential breaking of the linearisation approximations of the EKF.

3.5 Updating After a Measurement

After attempting measurements, those that were successful are used to update the state estimates. How do we know which were successful? Clearly there are some cases where failures are apparent, by matching scores below a given threshold for instance. However, in other cases we won't know: something has been found within the innovation covariance-bounded search region, but is it the feature we were looking for or just something else that looks the same?

One way to pick the good measurements from the bad is make a lot of measurements at the same time, and then look for sets among them which agree with each other: these are likely to be the correct matches, since no correlation is expected amongst the failures — this algorithm, RANSAC, is commonly used to lend robustness to batch methods [22]. To use RANSAC here, we try the update with randomly selected subsets of the measurements, and look for updated robot position estimates which agree. The bad matches can then be marked as such and the update performed with just the good ones.

To update the map based on a set of measurements z_i, we perform EKF updates as below. Because the measurements are independent, these updates can be done in sequence, rather than stacking all the measurements into one large vector and doing everything at once (this is computationally beneficial because smaller S matrices are inverted). Note further that if a particular z_i has diagonal measurement noise R we can further subdivide to the individual measurement parameters for sequential updates.

For each independent measurement $\mathbf{h}_i$, the Kalman gain $\mathtt{W}$ can be calculated and the state updated as follows:

$$\mathtt{W} = \mathtt{P}\frac{\partial \mathbf{h}_i}{\partial \mathbf{x}}^\top \mathtt{S}_i^{-1} = \begin{pmatrix} \mathtt{P}_{xx} \\ \mathtt{P}_{y_1 x} \\ \mathtt{P}_{y_2 x} \\ \vdots \end{pmatrix} \frac{\partial \mathbf{h}_i}{\partial \mathbf{x}_v}^\top \mathtt{S}_i^{-1} + \begin{pmatrix} \mathtt{P}_{xy_i} \\ \mathtt{P}_{y_1 y_i} \\ \mathtt{P}_{y_2 y_i} \\ \vdots \end{pmatrix} \frac{\partial \mathbf{h}_i}{\partial \mathbf{y}_i}^\top \mathtt{S}_i^{-1} \quad (4)$$

$$\hat{\mathbf{x}}_{new} = \hat{\mathbf{x}}_{old} + \mathtt{W}(\mathbf{z}_i - \mathbf{h}_i) \quad (5)$$

$$\mathtt{P}_{new} = \mathtt{P}_{old} - \mathtt{W}\mathtt{S}_i\mathtt{W}^\top . \quad (6)$$

3.6 Initialising a New Feature

When an unknown feature is first observed, a measurement $\mathbf{z}_i$ is obtained of its position relative to the robot. If the measurement function $\mathbf{h}_i(\mathbf{x}_v, \mathbf{y}_i)$ is invertible to $\mathbf{y}_i(\mathbf{x}_v, \mathbf{h}_i)$, we can initialise the feature state as follows (assuming here that two features are known and the new one becomes the third):

$$\mathbf{x}_{new} = \begin{pmatrix} \mathbf{x}_v \\ \mathbf{y}_1 \\ \mathbf{y}_2 \\ \mathbf{y}_i \end{pmatrix} \quad (7)$$

$$\mathtt{P}_{new} = \begin{bmatrix} \mathtt{P}_{xx} & \mathtt{P}_{xy_1} & \mathtt{P}_{xy_2} & \mathtt{P}_{xx}\frac{\partial \mathbf{y}_i}{\partial \mathbf{x}_v}^\top \\ \mathtt{P}_{y_1 x} & \mathtt{P}_{y_1 y_1} & \mathtt{P}_{y_1 y_2} & \mathtt{P}_{y_1 x}\frac{\partial \mathbf{y}_i}{\partial \mathbf{x}_v}^\top \\ \mathtt{P}_{y_2 x} & \mathtt{P}_{y_2 y_1} & \mathtt{P}_{y_2 y_2} & \mathtt{P}_{y_2 x}\frac{\partial \mathbf{y}_i}{\partial \mathbf{x}_v}^\top \\ \frac{\partial \mathbf{y}_i}{\partial \mathbf{x}_v}\mathtt{P}_{xx} & \frac{\partial \mathbf{y}_i}{\partial \mathbf{x}_v}\mathtt{P}_{xy_1} & \frac{\partial \mathbf{y}_i}{\partial \mathbf{x}_v}\mathtt{P}_{xy_2} & \frac{\partial \mathbf{y}_i}{\partial \mathbf{x}_v}\mathtt{P}_{xx}\frac{\partial \mathbf{y}_i}{\partial \mathbf{x}_v}^\top + \frac{\partial \mathbf{y}_i}{\partial \mathbf{h}_i}\mathtt{R}\frac{\partial \mathbf{y}_i}{\partial \mathbf{h}_i}^\top \end{bmatrix} \quad (8)$$

It should be noted that some bias is introduced into the map in initialising features in this way if (as is usual) the measurement process is non-linear.

If $\mathbf{h}_i$ is not invertible, it means that a single measurement does not give enough information to pinpoint the feature location (for instance a single view of a point feature from a single camera only defines a ray on which it lies). The approach that must be followed here is to initialise it into the map as a **partially initialised feature** $\mathbf{y}_{pi}$, with a different geometrical type (e.g. a line feature to represent the ray we know a point must lie on), and wait until another measurement from a different viewpoint allows resolution. At this stage a special second initialisation function $\mathbf{y}_i(\mathbf{x}_v, \mathbf{y}_{pi}, \mathbf{z}_i)$ allows the actual state $\mathbf{y}_i$ to be determined from the partially initialised state and new measurement (feature types which require more than two steps for initialisation are also possible).

Once initialised, a feature has exactly the same status in the map as those whose positions may have been give as prior knowledge.

3.7 Deleting a Feature

Deleting a feature from the state vector and covariance matrix is a simple case of removing the rows and columns which contain it. An example in a system where the second of three known features is deleted would be:

$$\begin{pmatrix} \mathbf{x}_v \\ \mathbf{y}_1 \\ \mathbf{y}_2 \\ \mathbf{y}_3 \end{pmatrix} \rightarrow \begin{pmatrix} \mathbf{x}_v \\ \mathbf{y}_1 \\ \mathbf{y}_3 \end{pmatrix} \;,\; \begin{bmatrix} P_{xx} & P_{xy_1} & P_{xy_2} & P_{xy_3} \\ P_{y_1 x} & P_{y_1 y_1} & P_{y_1 y_2} & P_{y_1 y_3} \\ P_{y_2 x} & P_{y_2 y_1} & P_{y_2 y_2} & P_{y_2 y_3} \\ P_{y_3 x} & P_{y_3 y_1} & P_{y_3 y_2} & P_{y_3 y_3} \end{bmatrix} \rightarrow \begin{bmatrix} P_{xx} & P_{xy_1} & P_{xy_3} \\ P_{y_1 x} & P_{y_1 y_1} & P_{y_1 y_3} \\ P_{y_3 x} & P_{y_3 y_1} & P_{y_3 y_3} \end{bmatrix} \;. \quad (9)$$

In automated map-maintenance, features can be deleted if a large proportion of attempted measurements fail when the feature is expected to be measurable. This could be due to features not fitting the assumptions of their model (an assumed point feature which in fact contains regions of different depths and therefore appears very different from a new viewpoint for instance), or possibly occlusion — leading to the survival of features which do not suffer these fates.

4 Software and Implementations

Realisation of the generic properties of the sequential map-building problem and experience with different robot systems has led to the evolution of our original application-specific software into a general framework called *Scene*, efficiently implemented in C++ and designed with orthogonal axes of flexibility in mind:

1. Use in many different application domains; from multiple robots navigating in 1D, 2D or 3D with arbitrary sensing capabilities, to single cameras.
2. Implementation of different mapping algorithms and approaches to dealing with the complexity of sequential map-building.

Scene is now available with full source code (under the GNU Lesser General Public License), at **http://www.robots.ox.ac.uk/~ajd/** . The distribution package includes interactive simulations precompiled for Linux which allow immediate hands-on experience of sequential map-building in several real and simplified problem domains, the additional tools which turn these into systems operating with real hardware, and substantial documentation.

To give an impression of how the general framework can be applied to various systems, details of some current and planned implementations, differentiated by motion and feature measurement models plugged in as modules, are given in Table 1. The simplest is a one-dimensional test-bed, which is very useful for looking at what happens to robot and feature covariances in various situations and under different algorithms. Our main work to date has been on robot navigation using active vision [5,7,6], using mobile platforms which move in 2D and are equipped with steerable camera platforms which make measurements of point features in 3D with stereo vision. As detailed in [6], the software's motion model formulation is flexible enough to permit cooperative position estimation

Table 1. Specifications for various implementations.

	1D Test-bed	Nomad Robot with Stereo Active Vision	Camera Position Tracking
World Dimension	1	3	3
Position Dimension	1	2	3
Motion Model	Velocity Only	Steer and Drive	General Motion in 3D
State Size	1	5	12
Control Size	1	3	0
Feature Measurement Model	Point Range Measurement	3D Stereo Point Measurement	Single Image Point Measurement
Feature Dimension	1	3	3
Measurement Size	1	4	2

by cooperating robots, where one has stereo active vision and the other is blind, navigating primarily by odometry.

Current PC computers are powerful enough to perform correlation searches for many features at video frame rate. Our current goal is to apply the *Scene* library to real-time camera position tracking using just inside-out image measurements, potentially the "killer app" of sequential localisation and map-building, which would be useful in applications such as inside-out head tracking or the real-time virtual studio. The first demonstrations of this type have just started to appear [3]. To be successful, it will be necessary to make use of many of the details explained in this paper. For instance, RANSAC or similar must be used to detect failed matches fast because there will not be enough processing time available to propagate multiple hypothesis. A full 3D motion model must be used, and finally, it will be necessary to use partially initialised feature representations to bootstrap features in 3D from multiple views.

Acknowledgements. The authors are grateful to Philip McLauchlan, Simon Julier and John Leonard for discussions.

References

1. P. A. Beardsley, I. D. Reid, A. Zisserman, and D. W. Murray. Active visual navigation using non-metric structure. In *Proceedings of the 5th International Conference on Computer Vision, Boston*, pages 58–65. IEEE Computer Society Press, 1995.
2. J. A. Castellanos. *Mobile Robot Localization and Map Building: A Multisensor Fusion Approach*. PhD thesis, Universidad de Zaragoza, Spain, 1998.
3. A. Chiuso, P. Favaro, H. Jin, and S. Soatto. "mfm": 3-d motion from 2-d motion causally integrated over time. In *Proceedings of the 6th European Conference on Computer Vision, Dublin*, 2000.
4. K. S. Chong and L. Kleeman. Feature-based mapping in real, large scale environments using an ultrasonic array. *International Journal of Robotics Research*, 18(2):3–19, January 1999.

5. A. J. Davison. *Mobile Robot Navigation Using Active Vision*. PhD thesis, University of Oxford, 1998. Available at http://www.robots.ox.ac.uk/~ajd/.
6. A. J. Davison and N. Kita. Active visual localisation for cooperating inspection robots. In *In Proceedings of the IEEE/RSJ Conference on Intelligent Robots and Systems*, 2000.
7. A. J. Davison and D. W. Murray. Mobile robot localisation using active vision. In *Proceedings of the 5th European Conference on Computer Vision, Freiburg*, pages 809–825, 1998.
8. H. F. Durrant-Whyte. Where am I? A tutorial on mobile vehicle localization. *Industrial Robot*, 21(2):11–16, 1994.
9. H. F. Durrant-Whyte, M. W. M. G. Dissanayake, and P. W. Gibbens. Toward deployments of large scale simultaneous localisation and map building (slam) systems. In *Proceedings of the 9th International Symposium of Robotics Research, Snowbird, Utah*, pages 121–127, 1999.
10. D. Fox, W. Burgard, H. Kruppa, and S. Thrun. Efficient multi-robot localization based on monte carlo approximation. In *Proceedings of the 9th International Symposium of Robotics Research, Snowbird, Utah*, pages 113–120, 1999.
11. C. G. Harris. Geometry from visual motion. In A. Blake and A. Yuille, editors, *Active Vision*. MIT Press, Cambridge, MA, 1992.
12. M. Isard and A. Blake. Contour tracking by stochastic propagation of conditional density. In *Proceedings of the 4th European Conference on Computer Vision, Cambridge*, pages 343–356, 1996.
13. E. T. Jaynes. Probability theory as extended logic. Technical report, Washington University in St. Louis, 1999. Web site: http://bayes.wustl.edu/.
14. Y. D. Kwon and J. S. Lee. A stochastic map building method for mobile robot using 2-d laser range finder. *Autonomous Robots*, 7:187–200, 1999.
15. J. J. Leonard and H. J. S. Feder. Decoupled stochastic mapping, part i: Theory. Preprint. Submitted to IEEE Transactions on Robotics and Automation, 1999.
16. P. F. McLauchlan and D. W. Murray. A unifying framework for structure and motion recovery from image sequences. In *Proceedings of the 5th International Conference on Computer Vision, Boston*. IEEE Computer Society Press, 1995.
17. M. Pollefeys, R. Koch, and L. Van Gool. Self-calibration and metric reconstruction in spite of varying and unknown internal camera parameters. In *Proceedings of the 6th International Conference on Computer Vision, Bombay*, pages 90–96, 1998.
18. R. Smith, M. Self, and P. Cheeseman. A stochastic map for uncertain spatial relationships. In *4th International Symposium on Robotics Research*, 1987.
19. S. Thrun, D. Fox, and W. Burgard. A probabilistic approach to concurrent mapping and localization for mobile robots. *Machine Learning*, 31, 1998.
20. P. H. S. Torr, A. W. Fitzgibbon, and A. Zisserman. Maintaining multiple motion model hypotheses over many views to recover matching and structure. In *Proceedings of the 6th International Conference on Computer Vision, Bombay*, pages 485–491, 1998.
21. P. H. S. Torr, R. Szeliski, and P. Anandan. An integrated bayesian approach to layer extraction from image sequences. In *Proceedings of the 7th International Conference on Computer Vision, Kerkyra*, pages 983–990, 1999.
22. P. H. S. Torr and A. Zisserman. Robust computation and parameterization of multiple view relations. In *Proceedings of the 6th International Conference on Computer Vision, Bombay*, pages 727–732, 1998.
23. J. K. Uhlmann, S. J. Julier, and M. Csorba. Nondivergent simultaneous map building and localisation using covariance intersection. In *The Proceedings of AeroSense:*

The 11th International Symposium on Aerospace/Defense Sensing, Simulation and Controls, Orlando, Florida. SPIE, 1997. Navigation and Control Technologies for Unmanned Systems II.

Discussion

1. **Daniel Cremers, University of Mannheim**: I have one question about the term "process noise". You mentioned that you think about it as process noise rather than random noise. I was wondering if that's just terminology or does it help you in solving the problems that you want to solve ?
    Andrew Davison: Maybe not, I think it's just something I realized. Noise accounts for the things you don't attempt to model in your system rather than actual random events. People always talk about noise and it makes you think of random things going on in the world but if you wanted to, you could model things better. For instance in our robot application you have process noise which represents the uncertainty in where the robot is after it's moved somewhere measuring its position by counting the number of times its wheels have turned. Something that leads to uncertainty in that situation is that the wheels sometimes slip on the floor. If we wanted to we could have a model of the floor, the wheel and the tyre and we could actually make that so that it wasn't uncertain but something that was modeled. We always model up to a certain level and then the rest we don't model, we choose not to. That's the role of process noise as I see it.

Panel Session on Extended Environments

Tomas Pajdla[1], J.P. Mellor[2], Camillo J. Taylor[3], Tomas Brodsky[4], and Andrew J. Davison[5]

[1] Center for Machine Perception, Dept. of Cybernetics, Faculty of Electrical Engineering, Czech Technical University, Prague, Czech Republic.
pajdla@cmp.felk.cvut.cz
[2] Rose-Human Institute of Technology, Terre Haute, IN 47803, USA.
j.p.mellor@rose-hulman.edu
[3] GRASP Laboratory, CIS Dept., Univ. of Pennsylvania, PA 19104-6229, USA
cjtaylor@central.cis.upenn.edu
[4] Philips Research, Briarcliff Manor, NY 10510, USA
Tomas.Brodsky@philips.com
[5] Intelligent Systems Division, Electrotechnical Laboratory, 1-1-4 Umezono, Tsukuba, Ibaraki 305-8568, Japan.
davison@etl.go.jp

1 Introduction

The topic of the third panel session was extended environments. Tomas Pajdla chaired the discussion and J.P.Mellor, C.J.Taylor, Tomas Brodsky and Andrew Davision also participated. Each panelist discussed the issues that he felt were going to be important for further developments in the modeling of extended environments. The panel session was followed by some questions and discussions which are also reported here.

2 Tomas Pajdla

Let me start the discussion about extended environments by saying a few words as the introduction. I will give my view. What are "extended environments"? We have seen examples of large outdoor environments or complex indoor environments (which we have not actually seen but which relates maybe to the last talk). We have also heard about some intrinsic problems, related to extended environments, like the reconstruction of large structures and building maps and navigation.

Actually it is interesting to see that there was a unifying theme about all the presentations. This was the use of certain non-classical cameras. In particular omnidirectional sensors. We saw the use of a catadioptric omnidirectional sensor, a special compound eye which can be also considered to be omnidirectional, and also the use of mosaicing which produces images that could be obtained from an omnidirectional sensor.

Having in mind these specifics of extended environments and the sensors which are used, we can ask which are the existing techniques which we should use and how to deal with extended environments using them. And then of course, what are new techniques which we don't know yet and also what are the main challenges?

I think that there are two important issues to address. First is error accumulation because the environment is large and we have many images, so we accumulate errors. We saw that there was a successful use of omnidirectional images in order to help stabilize ego-motion and camera localization. Of course it is a question of whether this is enough or if we still need some GPS.

Secondly, we need to work with a very large amount of imagery and complex models. So the question is if, for example, omnidirectional images can help us in this. They probably can but we still face the problem that we either have high frame-rates and low resolution (for catadioptric) or high resolution but low frame-rate (for mosaics).

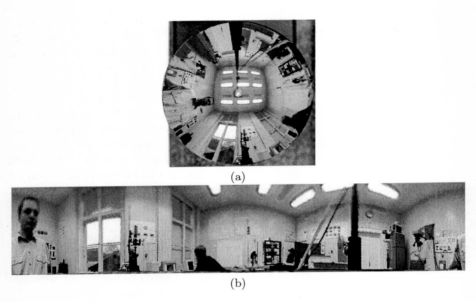

Fig. 1. (a) An original image of a mirror. (b) The warped image. Resolution in the upper part of the image (b) is lower because the upper part is transformed from the center of the image (a) where a small number of pixels covers a large view angle.

I would like to show two challenges which I believe are related to catadioptric sensors because this is my field and I can comment on that. One is related to a question which has been asked already: what about changing the resolution over the catadioptric sensor? Figure 1(a) shows an image of a mirror, taken by an ordinary CCD camera. In this case, in the center of the image, fewer pixels—which are in a square raster—cover the same view angle as do more pixels at the periphery. Therefore, in the center, we get rather low resolution compared

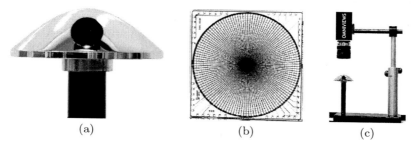

Fig. 2. A curved mirror (a) and a variant-resolution vision sensor (Courtesy of Giulio Sandini, Lira Lab, Genoa) (b) can provide even-resolution omnidirectional catadioptric cameras (c).

to the periphery as may be observed in the upper part of the warped image in Figure 1(b) and this, of course, may be not desirable. So the hope and the challenge is to build a sensor which will somehow combine a catadioptric camera with a special CMOS sensor with varying resolution and these two components may be matched together, as shown on Figure 2, so we get even resolution.

There is another thing I believe which may be seen as a challenge and this is to combine omnidirectional and ordinary images to make more complete reconstructions of environments. In order to do so, we have to model omnidirectional images correctly. The reason why they do not have the same model as normal images—and here of course we assume we have an omnidirectional camera with one center of projection—is that an omnidirectional image actually can't be modeled as a projective plane. In a projective plane, each ray is modeled by one line. However, in an omnidirectional image each line is covered by two opposite rays. Each ray may see a different point in space and therefore cannot be modeled by the same line. Another representation of an omnidirectional image, an oriented projective plane, is called for to do reconstructions from omnidirectional images correctly.

3   J. P. Mellor

I see four areas that I think will have a significant impact on how we approach extended environments. Some of these came up a bit earlier.

First, one of the most exciting developments are the new sources of data that are appearing. The omnidirectional image can be considered a new source of data that we are just now exploring. The work I presented uses both omnidirectional images (created by mosaicing individual images with the same center of projection) and GPS which is also becoming practical. Other interesting devices such as inertial sensors and Z-cam are getting small enough and cheap enough that they can be practically used. There are lots of exciting sensing technologies on the horizon and we should combine them with our vision work. They complement each other well.

Second is the amount of data versus the complexity of the system. This was mentioned briefly earlier. There is a trade-off between the amount of data and the complexity of the system (i.e. knowledge that you need to pour into it). The work that I did falls very much on the "large quantities of data, dumb systems" side. I'm not sure where we should be along this continuum, but I think we certainly need to explore the trade-offs.

Third is modeling. We heard, in Paul Debevec's work, some of the emphasis on reality. We also heard about the trade-off between geometry and image-based rendering. There are some interesting ways to do modeling that need to be explored. Perhaps a combination of geometric and procedural graphics. The graphics community has a lot to offer in this area.

Finally, systems. We need to put all these things together and get useful tools. David Nister is very interested in having a commercial product that the average person can just plug into his PC and use. A person who would know how to use a mouse, but probably doesn't know anything about computer vision. I think that if we just push in this direction, we'll get usable, robust systems and that's a good direction for us to be headed.

4 C. J. Taylor

Just two things I would like to mention:

The number one thing on my wish list is a decent real-time tracker for large environments. One of the things you see in extended environments is that they really push the state of the art of what you can do. We're not talking about a hundred frames, we're talking about thousands of frames. We're not talking about one foot, we're talking about a hundred feet. I think it could be interesting to see if we can extend our methods to deal with this.

The other thing that is important for immersive environments in particular is the issue of detail. What I did was to start of with a plenoptic approach. It's the easiest way to get the fine details. Of course, we would like to be able to do interpolation and fly around a bit. But I think in design and representations it is important to keep in mind that recovering and representing detail is probably going to be the most important part. People will live with large areas being modeled as flat walls, but if you miss some important details they start killing you. I think it will be interesting to go forward.

5 Tomas Brodsky

I'd like to mention several things. One, alternative models of space. The stratification into projective, affine and Euclidean spaces yielded many interesting results, but it is well known that humans do not build Euclidean models. For example, the visual space gets distorted by Cremona transformations. What might be interesting for the future is to study what models of space should be used.

Other things we want to look at are new camera models, similar to what I showed today. Especially if you synchronize camera networks, you can do many interesting things. For example, many people are interested in modeling dynamic human motion. Figure 3 shows some recent results from Maryland. In the input videos Brad Stuart shows off his martial arts skills. He's working on building dynamic models from many views. The first result uses space carving for each frame separately, but the second model in addition considers 3D motion fields and motion constraints in many views and you can see that the second model looks much better.

Fig. 3. Building dynamic models from many views

6 Andrew Davison

Maybe I can say something briefly about extended environments with respect to localization rather than map-building. If you're interested in, for instance, the position of a camera as it moves through the world, then, obviously, you're always going to be drifting away from your original coordinate frame if you're only ever measuring new features. For instance if you can see a certain number of features when you turn your system on and then when you move your camera a little bit, those features go out of view. Now your camera's position is somewhat uncertain and you initialize new features. The positions of those features are also going to be uncertain about the amount your camera is uncertain. By measuring those features as many times as you like, you'll never improve beyond that uncertainty. So you move on again and find some more new features. Your uncertainty is always going to grow. The only way you can reduce that is if you come back and look at the features you saw originally. Let's think about the system of head-tracking where you mount a camera on someone's head. If you just have a small field-of-view (normal camera), then if the person's going to move significantly, you're probably going to lose those features. I think it would be quite advantageous in that situation to use a wide-angle fish-eye camera or something like that, because from my work in active vision I've found that the big advantage active vision gives you is this huge field of view, you can match things over well over 180 degrees. You can see a feature, turn your robot 180 degrees and still see the same feature. So you can measure your rotation very accurately, which you couldn't do with a narrow field-of-view camera.

In active vision we get the best of both worlds. We can have very wide field-of-view and high resolution at every point because we just move our normal camera to look at each point. But if you had a fish-eye camera there's going to be some kind of trade-off. The disadvantage of active vision is that it's not efficient. You have to always move your cameras to look at different things whereas with a fish-eye camera you can look at everything at the same time. I think on the whole, the thing about being able to see features for longer is more important than the resolution issue.

Discussion

1. **Joss Knight, University of Oxford**: I wonder if there isn't a bigger gap between SFM and sequential localization than map building? I personally can not think of a huge number of applications in which you want to acquire dense structure models in real time. Usually what you want to do in real time is some kind of AI task. Mostly you want to localize yourself which we've established only needs a very small amount of data to do well. It doesn't need dense models and if you want to interact with something, you might need more dense information about your environment but not about things you're not currently interested in.
   Andrew Davison: I think that's a good point. I've always been more interested in localization. You can construct a kind of map maintenance criterion in which you're interested. For instance in my active vision localization: the robot should always be able to see at least two different features from where it is. If at a certain position it couldn't see two, then it would initialize a new one into its map. That's probably enough information for localization. Obviously it's a different problem if you try to build dense maps. The uncertainty information is still the same stuff but your focus is elsewhere. Maybe the big difference is that in localization you are potentially interested in extended times of driftless operation rather than extended environments.
   Joss Knight: I guess it's also worth pointing out that you have to remember that problems like getting pose for augmented reality is really a localization problem, not a dense SFM problem. Perhaps we try to use too much information for things like AR because all you really want to know is, say, the position of a couple of planes where you want to place your object and be well localized in terms of pose.
2. **Andrew Fitzgibbon, University of Oxford**: I think the question is more that if we're dealing with huge sequences, we have to deal with the issue of "forgetting". Andrew Davison is limited to, say, 100 features that can be tracked, and at some point he must forget some of his features in order to maintain constant speed. If you're going to take 20000 images, do you think

that the model should use all of those images or are you going to drop some, at some stage?

Andrew Davison: About forgetting, that's certainly important! You will always have an upper limit on the amount of features you can keep in your proper full covariance, good quality map. There have been many approaches suggested, like splitting up a big map in submaps. Then you can do good quality localization within that submap when you're in the area. You can keep the information about the other maps on some kind of back burner or you just store approximate information on how those submaps relate to each other.

C.J. Taylor: What I really want to do is solve the recognition problem instead of the reconstruction problem, the reason being that the environments we are dealing with are highly structured. It is really about recognizing that there are planes, that there are geometric surfaces. And if you could somehow get the system to do that automatically, that would solve your problem. You really start cutting down your number of parameters drastically. It would be great, tracking a wall rather than a cloud of points. When new features are coming out, you would test against this hypothesis.

3. **Marc Pollefeys**: I'd like to raise another issue: when you model large environments, some parts are not interesting and you can model them very roughly. Other parts are very interesting and you don't want to miss them and you want to model them on a higher resolution maybe. There are also issues about representation. You're not in a homogeneous representation anymore. Are there comments on this?

Tomas Brodsky: The question is: how do you recognize which parts are interesting?

Marc Pollefeys: Not only that. That's even harder if the computer has to recognize this. But what if you want to model a monument for instance and there is a main statue. Then you want to model this statue in a lot of detail, so you're going to acquire more data there. How to deal with that in your modeling efficiently and have this kind of representation which at some level is more accurate than at another ? I think these issues are very important for extended environments.

J. P. Mellor: You're right—it's very important. We assumed that somebody smart was taking the images. The density of the data affects the density of the sampling in the model that you get out. It would be very nice to have some automatic scheme for this. If the level-of-detail for a certain part was not sufficient, the system would say: go out and get some more data, more photographs.

Author Index

Aloimonos, Yiannis   204

Baker, Patrick   204
Brodsky, Tomas   235

Chai, Jinxiang   94
Cornelis, Kurt   144, 161
Culbertson, W. Bruce   109

Davison, Andrew J.   218, 235
Debevec, Paul   1, 161

Fermüller, Cornelia   204

Kanatani, Kenichi   35, 86
Kanazawa, Yasushi   35
Kang, Sing Bing   94, 161
Kita, Nobuyuki   218

Malzbender, Thomas   109
Mellor, J.P.   170, 235

Nistér, David   17, 86
Nobuyuki, Kita   218

Pajdla, Tomas   235
Pless, Robert   204
Pollefeys, Marc   144, 161
Ponce, Jean   52, 86

Shan, Ying   68
Shum, Heung-Yeung   94
Slabaugh, Gregory G.   109, 161

Taylor, Camillo J.   187, 235
Triggs, Bill   86

Van Gool, Luc   124, 144
Vergauwen, Maarten   144

Zalesny, Alexey   124
Zhang, Zhengyou   68, 86

Lecture Notes in Computer Science

For information about Vols. 1–1931
please contact your bookseller or Springer-Verlag

Vol. 1812: J. Wyatt, J. Demiris (Eds.), Advances in Robot Learning. Proceedings, 1999. VII, 165 pages. 2000. (Subseries LNAI).

Vol. 1932: Z.W. Raś, S. Ohsuga (Eds.), Foundations of Intelligent Systems. Proceedings, 2000. XII, 646 pages. (Subseries LNAI).

Vol. 1933: R.W. Brause, E. Hanisch (Eds.), Medical Data Analysis. Proceedings, 2000. XI, 316 pages. 2000.

Vol. 1934: J.S. White (Ed.), Envisioning Machine Translation in the Information Future. Proceedings, 2000. XV, 254 pages. 2000. (Subseries LNAI).

Vol. 1935: S.L. Delp, A.M. DiGioia, B. Jaramaz (Eds.), Medical Image Computing and Computer-Assisted Intervention – MICCAI 2000. Proceedings, 2000. XXV, 1250 pages. 2000.

Vol. 1936: P. Robertson, H. Shrobe, R. Laddaga (Eds.), Self-Adaptive Software. Proceedings, 2000. VIII, 249 pages. 2001.

Vol. 1937: R. Dieng, O. Corby (Eds.), Knowledge Engineering and Knowledge Management. Proceedings, 2000. XIII, 457 pages. 2000. (Subseries LNAI).

Vol. 1938: S. Rao, K.I. Sletta (Eds.), Next Generation Networks. Proceedings, 2000. XI, 392 pages. 2000.

Vol. 1939: A. Evans, S. Kent, B. Selic (Eds.), «UML» – The Unified Modeling Language. Proceedings, 2000. XIV, 572 pages. 2000.

Vol. 1940: M. Valero, K. Joe, M. Kitsuregawa, H. Tanaka (Eds.), High Performance Computing. Proceedings, 2000. XV, 595 pages. 2000.

Vol. 1941: A.K. Chhabra, D. Dori (Eds.), Graphics Recognition. Proceedings, 1999. XI, 346 pages. 2000.

Vol. 1942: H. Yasuda (Ed.), Active Networks. Proceedings, 2000. XI, 424 pages. 2000.

Vol. 1943: F. Koornneef, M. van der Meulen (Eds.), Computer Safety, Reliability and Security. Proceedings, 2000. X, 432 pages. 2000.

Vol. 1944: K.R. Dittrich, G. Guerrini, I. Merlo, M. Oliva, M.E. Rodriguez (Eds.), Objects and Databases. Proceedings, 2000. X, 199 pages. 2001.

Vol. 1945: W. Grieskamp, T. Santen, B. Stoddart (Eds.), Integrated Formal Methods. Proceedings, 2000. X, 441 pages. 2000.

Vol. 1946: P. Palanque, F. Paternò (Eds.), Interactive Systems. Proceedings, 2000. X, 251 pages. 2001.

Vol. 1947: T. Sørevik, F. Manne, R. Moe, A.H. Gebremedhin (Eds.), Applied Parallel Computing. Proceedings, 2000. XII, 400 pages. 2001.

Vol. 1948: T. Tan, Y. Shi, W. Gao (Eds.), Advances in Multimodal Interfaces – ICMI 2000. Proceedings, 2000. XVI, 678 pages. 2000.

Vol. 1949: R. Connor, A. Mendelzon (Eds.), Research Issues in Structured and Semistructured Database Programming. Proceedings, 1999. XII, 325 pages. 2000.

Vol. 1950: D. van Melkebeek, Randomness and Completeness in Computational Complexity. XV, 196 pages. 2000.

Vol. 1951: F. van der Linden (Ed.), Software Architectures for Product Families. Proceedings, 2000. VIII, 255 pages. 2000.

Vol. 1952: M.C. Monard, J. Simão Sichman (Eds.), Advances in Artificial Intelligence. Proceedings, 2000. XV, 498 pages. 2000. (Subseries LNAI).

Vol. 1953: G. Borgefors, I. Nyström, G. Sanniti di Baja (Eds.), Discrete Geometry for Computer Imagery. Proceedings, 2000. XI, 544 pages. 2000.

Vol. 1954: W.A. Hunt, Jr., S.D. Johnson (Eds.), Formal Methods in Computer-Aided Design. Proceedings, 2000. XI, 539 pages. 2000.

Vol. 1955: M. Parigot, A. Voronkov (Eds.), Logic for Programming and Automated Reasoning. Proceedings, 2000. XIII, 487 pages. 2000. (Subseries LNAI).

Vol. 1956: T. Coquand, P. Dybjer, B. Nordström, J. Smith (Eds.), Types for Proofs and Programs. Proceedings, 1999. VII, 195 pages. 2000.

Vol. 1957: P. Ciancarini, M. Wooldridge (Eds.), Agent-Oriented Software Engineering. Proceedings, 2000. X, 323 pages. 2001.

Vol. 1960: A. Ambler, S.B. Calo, G. Kar (Eds.), Services Management in Intelligent Networks. Proceedings, 2000. X, 259 pages. 2000.

Vol. 1961: J. He, M. Sato (Eds.), Advances in Computing Science – ASIAN 2000. Proceedings, 2000. X, 299 pages. 2000.

Vol. 1963: V. Hlaváč, K.G. Jeffery, J. Wiedermann (Eds.), SOFSEM 2000: Theory and Practice of Informatics. Proceedings, 2000. XI, 460 pages. 2000.

Vol. 1964: J. Malenfant, S. Moisan, A. Moreira (Eds.), Object-Oriented Technology. Proceedings, 2000. XI, 309 pages. 2000.

Vol. 1965: Ç. K. Koç, C. Paar (Eds.), Cryptographic Hardware and Embedded Systems – CHES 2000. Proceedings, 2000. XI, 355 pages. 2000.

Vol. 1966: S. Bhalla (Ed.), Databases in Networked Information Systems. Proceedings, 2000. VIII, 247 pages. 2000.

Vol. 1967: S. Arikawa, S. Morishita (Eds.), Discovery Science. Proceedings, 2000. XII, 332 pages. 2000. (Subseries LNAI).

Vol. 1968: H. Arimura, S. Jain, A. Sharma (Eds.), Algorithmic Learning Theory. Proceedings, 2000. XI, 335 pages. 2000. (Subseries LNAI).

Vol. 1969: D.T. Lee, S.-H. Teng (Eds.), Algorithms and Computation. Proceedings, 2000. XIV, 578 pages. 2000.

Vol. 1970: M. Valero, V.K. Prasanna, S. Vajapeyam (Eds.), High Performance Computing – HiPC 2000. Proceedings, 2000. XVIII, 568 pages. 2000.

Vol. 1971: R. Buyya, M. Baker (Eds.), Grid Computing – GRID 2000. Proceedings, 2000. XIV, 229 pages. 2000.

Vol. 1972: A. Omicini, R. Tolksdorf, F. Zambonelli (Eds.), Engineering Societies in the Agents World. Proceedings, 2000. IX, 143 pages. 2000. (Subseries LNAI).

Vol. 1973: J. Van den Bussche, V. Vianu (Eds.), Database Theory – ICDT 2001. Proceedings, 2001. X, 451 pages. 2001.

Vol. 1974: S. Kapoor, S. Prasad (Eds.), FST TCS 2000: Foundations of Software Technology and Theoretical Computer Science. Proceedings, 2000. XIII, 532 pages. 2000.

Vol. 1975: J. Pieprzyk, E. Okamoto, J. Seberry (Eds.), Information Security. Proceedings, 2000. X, 323 pages. 2000.

Vol. 1976: T. Okamoto (Ed.), Advances in Cryptology – ASIACRYPT 2000. Proceedings, 2000. XII, 630 pages. 2000.

Vol. 1977: B. Roy, E. Okamoto (Eds.), Progress in Cryptology – INDOCRYPT 2000. Proceedings, 2000. X, 295 pages. 2000.

Vol. 1978: B. Schneier (Ed.), Fast Software Encryption. Proceedings, 2000. VIII, 315 pages. 2001.

Vol. 1979: S. Moss, P. Davidsson (Eds.), Multi-Agent-Based Simulation. Proceedings, 2000. VIII, 267 pages. 2001. (Subseries LNAI).

Vol. 1983: K.S. Leung, L.-W. Chan, H. Meng (Eds.), Intelligent Data Engineering and Automated Learning – IDEAL 2000. Proceedings, 2000. XVI, 573 pages. 2000.

Vol. 1984: J. Marks (Ed.), Graph Drawing. Proceedings, 2001. XII, 419 pages. 2001.

Vol. 1985: J. Davidson, S.L. Min (Eds.), Languages, Compilers, and Tools for Embedded Systems. Proceedings, 2000. VIII, 221 pages. 2001.

Vol. 1987: K.-L. Tan, M.J. Franklin, J. C.-S. Lui (Eds.), Mobile Data Management. Proceedings, 2001. XIII, 289 pages. 2001.

Vol. 1988: L. Vulkov, J. Waśniewski, P. Yalamov (Eds.), Numerical Analysis and Its Applications. Proceedings, 2000. XIII, 782 pages. 2001.

Vol. 1989: M. Ajmone Marsan, A. Bianco (Eds.), Quality of Service in Multiservice IP Networks. Proceedings, 2001. XII, 440 pages. 2001.

Vol. 1990: I.V. Ramakrishnan (Ed.), Practical Aspects of Declarative Languages. Proceedings, 2001. VIII, 353 pages. 2001.

Vol. 1991: F. Dignum, C. Sierra (Eds.), Agent Mediated Electronic Commerce. VIII, 241 pages. 2001. (Subseries LNAI).

Vol. 1992: K. Kim (Ed.), Public Key Cryptography. Proceedings, 2001. XI, 423 pages. 2001.

Vol. 1993: E. Zitzler, K. Deb, L. Thiele, C.A.Coello Coello, D. Corne (Eds.), Evolutionary Multi-Criterion Optimization. Proceedings, 2001. XIII, 712 pages. 2001.

Vol. 1995: M. Sloman, J. Lobo, E.C. Lupu (Eds.), Policies for Distributed Systems and Networks. Proceedings, 2001. X, 263 pages. 2001.

Vol. 1997: D. Suciu, G. Vossen (Eds.), The World Wide Web and Databases. Proceedings, 2000. XII, 275 pages. 2001.

Vol. 1998: R. Klette, S. Peleg, G. Sommer (Eds.), Robot Vision. Proceedings, 2001. IX, 285 pages. 2001.

Vol. 1999: W. Emmerich, S. Tai (Eds.), Engineering Distributed Objects. Proceedings, 2000. VIII, 271 pages. 2001.

Vol. 2000: R. Wilhelm (Ed.), Informatics: 10 Years Back, 10 Years Ahead. IX, 369 pages. 2001.

Vol. 2003: F. Dignum, U. Cortés (Eds.), Agent Mediated Electronic Commerce III. XII, 193 pages. 2001. (Subseries LNAI).

Vol. 2004: A. Gelbukh (Ed.), Computational Linguistics and Intelligent Text Processing. Proceedings, 2001. XII, 528 pages. 2001.

Vol. 2006: R. Dunke, A. Abran (Eds.), New Approaches in Software Measurement. Proceedings, 2000. VIII, 245 pages. 2001.

Vol. 2007: J.F. Roddick, K. Hornsby (Eds.), Temporal, Spatial, and Spatio-Temporal Data Mining. Proceedings, 2000. VII, 165 pages. 2001. (Subseries LNAI).

Vol. 2009: H. Federrath (Ed.), Designing Privacy Enhancing Technologies. Proceedings, 2000. X, 231 pages. 2001.

Vol. 2010: A. Ferreira, H. Reichel (Eds.), STACS 2001. Proceedings, 2001. XV, 576 pages. 2001.

Vol. 2013: S. Singh, N. Murshed, W. Kropatsch (Eds.), Advances in Pattern Recognition – ICAPR 2001. Proceedings, 2001. XIV, 476 pages. 2001.

Vol. 2015: D. Won (Ed.), Information Security and Cryptology – ICISC 2000. Proceedings, 2000. X, 261 pages. 2001.

Vol. 2018: M. Pollefeys, L. Van Gool, A. Zisserman, A. Fitzgibbon (Eds.), 3D Structure from Images – SMILE 2000. Proceedings, 2000. X, 243 pages. 2001.

Vol. 2021: J. N. Oliveira, P. Zave (Eds.), FME 2001: Formal Methods for Increasing Software Productivity. Proceedings, 2001. XIII, 629 pages. 2001.

Vol. 2024: H. Kuchen, K. Ueda (Eds.), Functional and Logic Programming. Proceedings, 2001. X, 391 pages. 2001.

Vol. 2027: R. Wilhelm (Ed.), Compiler Construction. Proceedings, 2001. XI, 371 pages. 2001.

Vol. 2028: D. Sands (Ed.), Programming Languages and Systems. Proceedings, 2001. XIII, 433 pages. 2001.

Vol. 2029: H. Hussmann (Ed.), Fundamental Approaches to Software Engineering. Proceedings, 2001. XIII, 349 pages. 2001.

Vol. 2030: F. Honsell, M. Miculan (Eds.), Foundations of Software Science and Computation Structures. Proceedings, 2001. XII, 413 pages. 2001.

Vol. 2031: T. Margaria, W. Yi (Eds.), Tools and Algorithms for the Construction and Analysis of Systems. Proceedings, 2001. XIV, 588 pages. 2001.

Vol. 2034: M.D. Di Benedetto, A. Sangiovanni-Vincentelli (Eds.), Hybrid Systems: Computation and Control. Proceedings, 2001. XIV, 516 pages. 2001.